Dunes

Dunes

Dynamics, Morphology, History

Andrew Warren

WILEY-BLACKWELL

A John Wiley & Sons, Ltd., Publication

This edition first published 2013
© 2013 John Wiley & Sons, Ltd

Wiley-Blackwell is an imprint of John Wiley & Sons, formed by the merger of Wiley's global
Scientific, Technical and Medical business with Blackwell Publishing.

Registered Office
John Wiley & Sons, Ltd, The Atrium, Southern Gate, Chichester, West Sussex, PO19 8SQ, UK

Editorial Offices
350 Main Street, Malden, MA 02148-5020, USA
9600 Garsington Road, Oxford, OX4 2DQ, UK
The Atrium, Southern Gate, Chichester, West Sussex, PO19 8SQ, UK

For details of our global editorial offices, for customer services, and for information about how
to apply for permission to reuse the copyright material in this book please see our website at
www.wiley.com/wiley-blackwell.

The right of Andrew Warren to be identified as the author of this work has been asserted in
accordance with the UK Copyright, Designs and Patents Act 1988.

Library of Congress Cataloging-in-Publication Data

Warren, Andrew.
 Dunes : dynamics, morphology, history / Andrew Warren.
   pages   cm
 Includes bibliographical references and index.
 ISBN 978-1-4443-3969-7 (cloth) – ISBN 978-1-4443-3968-0 (pbk.)
 1. Sand dunes–History.   2. Geomorphology.   3. Sand dunes–Environmental aspects.
 GB631.W36 2013
 551.3′75–dc23

                                   2012050118

A catalogue record for this book is available from the British Library.

Cover image: Sand dune © Isabella Pfenninger / iStockphoto
Cover design by Workhaus

Set in 10/12pt Plantin by SPi Publisher Services, Pondicherry, India
Printed in Malaysia by Ho Printing (M) Sdn Bhd

1   2013

Contents

List of Figures

Acknowledgements

I am deeply indebted to: Miles Irving at UCL, for his excellent draftsmanship, his attention to detail, his innovation and his forbearance of my many changes of mind; the Department of Geography at UCL, for allowing Miles to work with me and for allowing me access to the library at UCL; Joanna Bullard, whose eagle eye spotted many faults in the first-submitted draft of this book, and who made many valuable comments; my wife, for her constructive indifference; my family, who made a joke of the long gestation of this book; Ian Wilson, who thought big about dunes; my research students Adrian Chappell, Giles Wiggs and Hiroshi Momiji, whose work provided some of the most stimulating material reviewed here and from whose many publications I have extensively borrowed; John Stout in Big Spring, Texas, who has collated the vast and immensely useful Bibliography of Aeolian Research; and my erstwhile co-author, Ian Livingstone (our book was the groundwork for this one); Mary Bourke (Trinity College Dublin) for help with images of Mars; Paul Hesse (University of Wollongong), for the use of a layer of his GIS of Australian dunes; and Peter Bull at Oxford for the photomicrographs of grains of sand; Jack Gillies (Desert Research Institute, Nevada) for his help in getting me a high-enough resolution version of Figure 6.1.

Andrew Warren
London, January 2013

Introduction

This book is a search for explanations of the form, dynamics, size, pattern and history of windblown dunes. Any such search, even today, must begin with Bagnold's (1941) classic, *The Physics of Blown Sand and Desert Dunes*, a work that has been cited in almost every subsequent publication on dunes.

Beyond registering 34 more citations, and thus paying at least as many compliments, my purpose is to make two complements to *Blown Sand*. The first is a review of the research that it has inspired. The second is to add accounts of dunes formed round living plants, on coasts, in the past and in managed environments. Despite their absence from *Blown Sand*, the explanation of many of these other dunes is almost as dependent on its insights as the dunes it did cover. I have not gone to the extent of including dune-like features on the beds of rivers, estuaries and seas, even: when they share many characteristics of their shape and dynamics; when the word 'dune' is now attached to them in the scientific literature; and when *Blown Sand* has lessons for them (as Bagnold later confirmed). I justify this decision, first by the exclusion of these other dunes from the common understanding of the word 'dune'; second, by noting that the contrast in the density of wind and water is responsible for a significant difference between their dunes, such that subaqueous dunes in the same size-range as windblown dunes are the dynamic equivalent to windblown mega-dunes, while subaqueous ripples are the dynamic equivalent of windblown dunes (Chapter 3); third, by the small number of places where I have had to invoke subaqueous dunes; and fourth, and finally, most realistically, because the inclusion of subaqueous dunes would have significantly enlarged this book.

Dunes: Dynamics, Morphology, History, First Edition. Andrew Warren.
© 2013 John Wiley & Sons, Ltd. Published 2013 by John Wiley & Sons, Ltd.

The surge of publications on windblown dunes that began in the 1950s (Stout *et al.* 2009) has been maintained in the last two decades by an expansion in the national origins of the research, most evident in the near-dominance of the literature in some years by authors in China, India and South America. The surge has also been driven by an equally healthy disciplinary expansion. The range of skills now focused onto dunes includes: mathematical modelling of many kinds; vastly improved instrumentation for the measurement of sand movement in the field and in the wind tunnel; steadily more sophisticated analysis of increasingly better resolved remotely sensed images; dating techniques (especially optically stimulated luminescence, Chapter 10); stratigraphy; pedology; and others. Few can claim to have mastered all these fields, and I have struggled with some, but not to the extent, I hope, that I have not given them their correct weight. My account will show that there are some very active areas of research, such as the dating of palaeodunes, the measurement of flow over dunes, the modelling of dune profiles and patterns, and models of linear dune formation, where my judgements will quickly and surely be overtaken.

The organisation of this book follows the seminal system of the late Stanley Schumm and his colleague, Robert Lichty, which linked the size, age and complexity of landforms (Schumm and Lichty, 1965, RL; references in the text marked 'RL' are listed in the References at the end of each chapter). With one significant exception, their insight applies very well to dunes. As the size of a dune (ripple or group of dunes) expands, the longer it takes to develop, the more complex must be its explanation (as more processes are involved), and its primary explanatory variable may change. The study of windblown sand even adds dimensional expansion to the Schumm-and-Lichty framework. Thus, Part I of this book is about the smallest relevant space/timescale ($<10\,m^2$; <10 years), namely the lifting and carrying of sand by the wind (Chapter 1), ripples (Chapter 2), and the cross-sectional form of a single transverse dune (Chapter 3), in all of which the wind comes from one direction only, and where there is an unlimited stock of sand, in other words a two-dimensional frame, in which the critical variables are more or less confined to only wind speed, grain size and moisture content. Part II ($1000–10,000\,m^2$; $100–1000$ years) must bring in the third spatial dimension if it is to make sense of dune patterns (Chapters 4–7). In this larger frame, two new variables become critical, namely variability in the direction of the wind and the availability of sand. For this change in spatial scale, there must be, following Schumm and Lichty, an expansion in the temporal scale. Part III contains the significant exception to the Schumm-and-Lichty scheme: the space scale must start at 0.3 mm (the approximate size of a grain of dune sand) to allow explanations of the shape and surface texture of single grains as they have evolved over thousands of years of being blown about, while the timescale expands to 2200 million years, that being the likely time since the appearance on Earth of the primordial dune. This space-time frame allows explanations of sand seas (Chapter 8); the history of blown sand (Chapter 9); windblown sandstones and continental palaeodunes (Chapter 10); palaeocoastal dunes

(Chapter 11); and, finally, dunes on Mars, Venus and Titan (Chapter 12). Part IV (Care) returns to a neater space-time frame: first, Local Care in the Short Term (Chapter 13; <1000 m²; <10 years); and second in the long term, in which the aim is Sustainability (Chapter 14; >100,000 m²; >10 years).

The coverage of all this material could be at a number also of explanatory scales, ranging from mere stimulation, to lengthy, rigorous examination of models and the evidence by which they are to be judged. I have taken the broad middle road. It is the path of a geomorphologist, with field experience of aeolian processes, aeolian morphologies and the history of windblown landforms, and delivers a book of manageable size. The target audience is geomorphologists, perhaps not those at school or in the early years at university, but those in the later years and beyond, and for those interested in the geomorphology of related landforms, and of other disciplines, such as ecology, archaeology, geology and engineering, in which dunes play a role.

I have taken the opportunity that Google Earth offers to cite the latitudes and longitudes and 'eye altitudes' (in metres or kilometres) of scenes that help the argument. However, a caution about some of the scenes at low eye altitudes: over the years of writing this book, I have noticed that Google Earth frequently updates its images, so that, in some cases, the message I intended to be taken from an image has been lost, as where dunes show unmistakable evidence of a change in wind direction, or where there have been significant changes in the profiles of sand-yielding beaches. Google Mars also helps in understanding Chapter 12. Another website that takes much of the burden of explanation is Wikipedia, where explanations can be found of many of the terms and concepts not fully explained in the text.

Most of the references in the text are to the most recent paper to cover the topic in question (not necessarily the pioneer publication). In most cases, there are many predecessors, most of which are listed in the Bibliography of Aeolian Research (BAR) (http://www.csrl.ars.usda.gov/wewc/biblio/bar.htm). The great majority of references in the book are from the BAR, and therefore need no listing here. The cross-references 'earlier' and 'later' in a chapter of what follows refer to other parts of the same chapter. References in the text to the BAR conform to the following system: the letters 'a', 'b', 'c', etc. after the date of a reference refer to the order in which the same family name is listed in the BAR, not necessarily to the author in the reference (for example, a citation of 'Smith, 2000c' refers to the third Smith in the BAR list, even if the other Smiths are not cited in this book). Chinese names, where possible, are in the format 'Xi Jinpeng', and are listed under the family name (e.g. 'Xi') first. The BAR lists single-author papers first, then two-author papers, then papers by three authors or more.

Reference

Schumm, S.A. and Lichty, R.W. (1965) 'Time space and causality in geomorphology', *American Journal of Science* 263 (2): 110–119.

Part One
$<10 \text{ m}^2$; <10 years

Chapter One
Wind and Sand

Part I of this book (in which this is the first chapter) has a two-dimensional spatial frame. The principal consequence is that the wind is assumed to come from one direction only.

The chapter is about the lifting and carrying of sand by the wind. As to winds at this scale, only their velocity and small-scale patterns of turbulence are relevant. As to sand, the primary interest here is in its grain size (shape in a secondary concern). In the most widely accepted taxonomy for the size of sedimentary particles, grains of 'sand' have diameters of between 0.625 mm and 2 mm (Wentworth, 1922, RL). A global survey, by its own admission limited, but probably representative, found that most dune sands were in Wentworth's 'fine sand' category (0.10–0.40 mm) (Ahlbrandt, 1979). A quick scan of recent papers confirms this. Another assumption here is that sand is composed only of quartz. Dunes built of smaller and coarser particles, and of other minerals, are described in Chapter 6. Because changes in the shape and surface texture of windblown sand grains have long histories, they are issues for Chapter 4.

Wind versus Bed

The mechanical energy spent when two bodies (such as the wind and the surface) slide past each other is termed 'shear'. Shear on a surface over which a wind passes is denoted by τ_0.

Dunes: Dynamics, Morphology, History, First Edition. Andrew Warren.
© 2013 John Wiley & Sons, Ltd. Published 2013 by John Wiley & Sons, Ltd.

The Law of the Wall

Because, until very recently, the shear of the wind on a surface could not be measured, it had to be approached through theory. The first model of the relationship grew out of the work of Ludwig Prandtl and Theodore von Kármán, working first at Göttingen, and came to be known as the 'Law of the Wall', where the word 'wall' was chosen with wind tunnels in mind. For dunes, 'bed' (as in the 'bed of a stream') is more appropriate than 'wall' and is adopted here. Despite some serious revision (shortly), the Law is still a good introduction to how the wind shears a surface.

The Law is built on two observations (or, as will be seen, simplifying assumptions):

- The velocity of the wind increases upward from the bed, because friction on the bed retards the wind, and this retardation is transferred, with weakening effect, to the wind at greater heights. Figure 1.1a shows the velocity of a wind at successive heights above the bed of a wind tunnel, both on arithmetic scales. In Figure 1.1b, the data and the velocity scale are the same as in Figure 1.1a, but the height scale is now logarithmic. The Law declares that on a semi-logarithmic plot, like this, the data fall on a straight line, as they do in this and many other observations in wind tunnels. The slope of the straight line in Figure 1.1b is a function of the strength of the wind that it represents: steep slopes represent gentle winds (low velocities even at some height); gentler slopes represent faster, more powerful winds (high velocities near the bed). This relationship comes into the argument again shortly.
- Figure 1.1b also shows that the straight lines depicting winds of different strength reach the same focal point on the vertical (height) axis. Those who built the Law took this to imply that there was a very thin layer of air, just above the bed, at the same height for all winds, where the air was stationary or moving very slowly. The focal point is higher on rougher beds, which is why it is termed the 'roughness height' or 'roughness length' (shorthand, 'z_0'). Because few measuring instruments, even today, can penetrate this layer, z_0 has to be estimated. A common estimate is ~1 mm over a smooth, stable sandy surface on Earth. A newer formulation of z_0 is discussed very shortly.

In order for the Law to apply to fluids of differing density, a parameter, shear velocity (or friction velocity), or u_*, was introduced:

$$u_* = \sqrt{\tau_0/\rho_a} \, ,$$

where τ_0 is the shear force on the bed, and ρ_a is the density of the fluid; this is the 'Prandtl–Kármán equation'.

Because the dimensions of u_* are those of velocity ($m\,s^{-1}$), it is termed the 'shear velocity'. The equation is applicable to thin air on the Tibetan plateau, or on

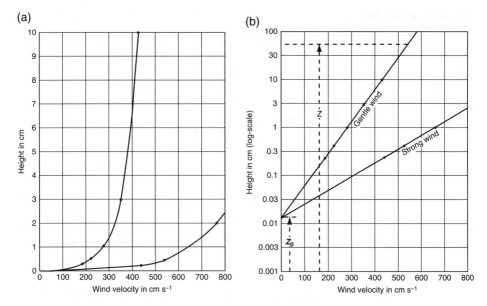

Figure 1.1 Velocity profile of the wind over a smooth, flat bed in a wind tunnel. In (a), both axes are arithmetic; in (b), the velocity (on the abscissa) is still arithmetic, but the height scale (on the ordinate) is now logarithmic. The data are Bagnold's (1941, pp. 48–49). The notations are explained in the text. Plots like this are known as 'Prandtl curves'. Redrawn after Bagnold (1941).

Mars; to dense atmospheres at very low temperatures, as near the Poles on Earth, or on Venus; or to any denser fluid, including water. The shorthand, 'u_*', is used in many places in this and later chapters.

The next step in the building of the Law was to relate u_* to measurements of the slope of the velocity/log height plot (Figure 1.1b). This step built on the observation (earlier) that the slopes of the lines are related to the strength of the wind. More precisely,

$$u_z / u_* = 1 / \kappa \ln\left(z / z_0\right),$$

where u_z is wind velocity at height z, and κ is Kármán's constant.

There are various estimates of κ, from theory and experiment; most are ~0.40.

Improving the wind/bed model

For all its simplicity and clarity, the Law is now only the first step in understanding wind shear on a solid surface. Three of its essential components have been found to be oversimplifications.

The velocity profile

Many plots of the velocity of flow against height over a bed do not fit the logarithmic profile. This is shown by attempts to fit straight lines to measurements of the wind at different heights above a beach, which produced a wide range of values of z_0 (Bauer *et al.*, 1992). Similar problems have been found with measurements in a wind tunnel (Butterfield, 1999a). The implication is that the way in which u_* and z_0 in the Law were derived was somewhat arbitrary.

There are many reasons for divergence from the semi-logarithmic curve. Three apply to common situations on dunes.

- When sand is in motion, there is a change in the slope in the velocity/height profile near the top of the cloud of bouncing sand. This point has been dubbed 'Bagnold's kink', after its discoverer (Bagnold, 1941, p. 59) (McEwan, 1993).
- Above a heated surface, which is the norm in deserts during the day, the air is unstable, and the wind-velocity profile, below about 0.5 m from the bed, is not semi-logarithmic. Errors of up to 15 times the value of u_* may occur if the stability condition of the atmosphere is not allowed for (Frank and Kocurek, 1994). The reason for this kind of deviation in the wind-velocity profile is that heat then joins shear as a driver of vertical mixing, and ensures that there is less change in velocity with height than in neutral conditions. The sand-flow rate in these conditions reaches equilibrium more quickly than in neutral conditions (Lu Ping and Dong ZhiBao, 2008). In very stable conditions, as during cold nights in some deserts and, more distinctly, in extreme cold at high latitudes or altitudes, the profile also deviates from the logarithmic. Mosaics of surfaces with different responses to heating, also the norm in deserts, create even more complex wind-velocity profiles as the wind repeatedly passes over surfaces of different roughness (Butler *et al.*, 2005).
- The height distribution of velocity above a sloping or curved bed is less easy to accommodate in estimates of shear. This is an issue in the explanation for flow over dunes (Chapter 3).

There are now many alternatives to the semi-logarithmic model of the relation between wind velocity and height, but most are useful only in wind tunnels, and few have been used by geomorphologists (Bauer *et al.*, 2004).

The roughness height (z_0)

This was the second element in the Law to be questioned. As explained earlier, the definition of z_0 in the Law is as arbitrary as that of u_*.

The most commonly used estimate of z_0 in a cloud of bouncing sand is now Owen's model (1964):

$$z_0 = \alpha\left(u_*^2 / 2g\right),$$

where g, as ever, is the gravitational constant.

Owen proposed that $\alpha = \sim 2$; others have found different values (Rasmussen *et al.*, 1996). Owen's model assumes that the cloud of salting sand is a form of roughness. Unlike z_0 in the Law, z_0 in Owen's model is not at the same height in all winds. Owen also discovered that shear on a bed under a cloud of saltating sand was the same as the shear anywhere within the saltating cloud. In other words, the saltating cloud is a self-organising system, the first of a number of self-organising systems that shape dunes, as will be explained later, especially in Chapter 4. As with the height/velocity curve, there are now several other formulations of z_0 for situations in which sand is in movement (detail in Bauer *et al.*, 2004).

Microturbulence

Turbulence is one of the main ways in which energy is transferred from the wind to the bed. In the scale limits of this chapter, it is only small-scale turbulence that is relevant. At this scale, turbulence is structured into 'burst–sweep' sequences. A burst is a downward spurt of air that replaces the air that has just been removed by a sweep, which is a slower upward ejection from the bed. On a loose sandy bed, the leading edge of a burst may dislodge sand, which is then taken up by a sweep. Burst–sweep sequences are responsible for most entrainment, even when, as is probably common, they are effective for only about 20% of the time (Sterk *et al.*, 1998). Measurements on a sandy, eroding field in Burkina Faso revealed that the burst–sweep sequence (at that site and on that occasion) had a downwind dimension of 0.25 m (Leenders *et al.*, 2005).

Turbulence at this scale can be measured by the Reynolds stresses, which describe the variation of velocity in three dimensions. Velocity in the windward direction is denoted 'u'; in the vertical (up or down) 'w'; and in the lateral (sideways in either direction) 'v'. u is positive downwind; w is positive upward; v would take the discussion beyond the two-dimensional frame of this chapter; its role in determining the two-dimension form of dunes has anyway hardly been explored. In a burst (towards the bed), u' is positive (wind-directional flow faster than the mean), and w' is negative (flow more downward than the mean). In a sweep (movement away from the bed), u' is negative (wind-directional flow in the sweep being slower than the mean), and w' is positive (upward flow at a faster rate than the mean). The prime symbol (') denotes a fluctuating variable.

The best measure of shear on the bed

This is the most important issue raised by all these reservations about the Law. Shear velocity (u_*) is, by definition, a description of the mean wind, which is seen in the relationship between u_* and the transport rate at progressively smaller averaging intervals (Namikas *et al.*, 2003). Thus, any study of entrainment at a small scale must acknowledge burst–sweep sequences. One alternative for measuring turbulent flow is the 'Reynolds Shear Stress', which combines stresses in the forward and vertical dimensions: $-\overline{u'w'}$ (the overbar denotes the mean). When multiplied by the air density (ρ) this gives a force, $\rho\left(-\overline{u'w'}\right)$.

The value of u_* has also been questioned even at larger scales. The solution could be as simple as deriving u_* from a wind profile measured down to about 0.05 m of the bed (Bauer *et al.*, 2004), or measuring the free-stream velocity 'well above' the bed (Schönfeldt and von Löwis, 2003), both of which solutions are empirical rather than theoretical. Developments in measuring sand transport, including shear or force balances, may deliver more theoretically acceptable measures of shear (Gillies *et al.*, 2000).

Lift-Off

A particle of sand starts to move (is 'entrained') when the forces that hold it down are exceeded by those that might rip it away.

Holding down by gravity

The gravitational force is defined thus:

$$g\left(\rho_{\mathrm{p}} - \rho_{\mathrm{a}}\right)d^3,$$

where g is the acceleration owing to gravity; ρ_{p} is particle density; ρ_{a} is the density of the air (or other fluid); and d is the particle diameter.

In other words, where the densities (ρ_{p} and ρ_{a}) are constant (when all the sand is of the same mineral, say quartz, and the air density does not change, as in many situations), particles of greater size (d), are held down by a greater gravitational force. The model can accommodate the behaviour of sands that are denser than quartz (say, of magnetite) or less dense (say, of diatomite), and of fluids with different density. Chapter 12 includes a discussion of the effects of the differences in gravity, temperature and air density on the lifting and carrying of sand by the wind on Mars, Venus and Titan.

Holding down by cohesion

Cohesion derives from several of the characteristics of particles. First, finer particles pack more closely, which means that they touch each other in more places and are thus more coherent. Second, rough particles touch each other in many more places than do smooth ones and so also cohere better. Third, platy shapes, as of many fine particles (particularly clays) allow much more contact than do rounded shapes, if packing is parallel (as it usually is for clays). Fourth, physicochemical bonds, known as London–van der Waals forces, increase cohesion between clay particles of some mineralogies (many clays) but are weaker between particles of some common rock minerals, such as quartz (Cornelis and Gabriels, 2004).

The fifth and sixth forms of cohesion come from water held between particles. The fifth is the meniscus force, which is strongest where a meniscus has a small angle of contact with a particle. This is the case where the voids between particles are small, as they are between fine particles. The strength of this force also depends on the roughness, roundness and surface properties of the particles, and on contaminants in the water.

None of the cohesive forces is as strong as the sixth form of cohesion. It depends on water held (adsorbed) on the surfaces of particles. The amount of water held in this way increases with relative atmospheric humidity, but, contrary to intuition, the static threshold (shortly) peaks at a relative humidity of 35–40%, below and above which value, entrainment is easier (Ravi and D'Orico, 2005). All of these properties are difficult to measure, and their combined effect is a major challenge to modellers and experimentalists (McKenna-Neuman and Sanderson, 2008). The relation between moisture and movement is discussed again later in the section on the dynamic behaviour of moisture in an eroding bed.

In sum, fine particles are more coherent than coarse, other things being equal, which they often are. Cohesion, of all sorts, is a function of $(1/d)^3$, where d is the diameter of the particles. This is the start of an explanation for why dunes are sandy: fine particles cohere too well to be easily moved by the wind (with some exceptions, later).

Raising by lift

Shear moves particles on a loose bed by two kinds of 'aerodynamic entrainment'. The first, lift, occurs because fast flow is accompanied by low pressure, following Bernoulli's equation. Flow over a bed of particles is faster than the velocity of the surrounding fluid in two situations: first, where there is a difference in pressure between the slow flow very near the bed, and the faster flow just above it, the slope of the velocity/height curve being steepest there; and second, where the wind is accelerated over a protruding particle.

Lift is more effective on rough than on smooth beds, on moving than static particles, and in sudden changes in velocity, as caused by turbulence. In some circumstances, lift may also be augmented by thermal diffusion from a heated surface, or by electrostatic forces arising from friction between the wind and the sediment (Rasmussen *et al.*, 2009). Lift alone entrains few particles, but it lightens the task of the other forces.

Raising by drag

Drag is usually stronger than lift. 'Surface drag' is caused by friction between the wind and the bed. It causes both rolling and sliding. 'Form drag' is caused by the difference in pressure between the windward and lee sides of a protruding particle, especially in high turbulence. Because it is greatest on top of the particle, it causes rolling. Both contribute to 'aerodynamic moment', which is the force on a particle that is dependent on its projection above the surface: particles that project more (bigger or longer ones) are toppled and therefore entrained more easily. Drag can help to eject particles when they collide or are dragged over projections. The magnitude of both forms of drag is proportional to $u_*^2 d^2$ (d being the particle diameter). Drag, like lift, is effective only very close to the bed, and raises few grains on its own. When the particles are clear of the bed, many acquire spin, which may contribute up to 24% of lift (the Magnus effect) (Huang Ning *et al.*, 2010).

Raising by bombardment

When particles are lifted into the wind, they pick up momentum from the wind in their trajectory above the surface and take some back to the bed on their return. This is 'bombardment', and, when sand is in movement, it is more powerful than any of the other sand-raising mechanisms. It both dislodges loose grains and breaks up aggregates (pellets) and crusts (both later). Once sand is lifted in sufficient amounts, further entrainment is almost wholly by bombardment.

Thresholds

This section adds to, but does not yet complete, the explanation for why dunes are built of sand. The wind can lift grains if it has sufficient power, which is to say (following the line of reasoning earlier), it is fast enough. The critical condition, when sand begins to blow, is the threshold velocity (u_t), or, more generally, the threshold shear velocity (u_{*t}).

In Bagnold's (1941, pp. 85–90) terms, the *static* (or fluid) threshold is the wind velocity at which grains start to move under the influence of lift and drag alone; and the dynamic (or impact) threshold is passed when particles are bombarding the bed. The dynamic threshold occurs at a lower velocity than the static threshold, because of bombardment. A much later model shows that in most places on Earth, the velocity at the dynamic threshold is ~0.96 (the ratio is different on Mars, Venus and Titan, as described in Chapter 12; Almeida *et al.*, 2008a).

Bagnold (1941, pp. 85–90) built the first mathematical model of the static threshold, which described the balance between the lifting and retaining forces on a particle. Subsequent theorising was reviewed by Cornelis and colleagues (2004b; Figure 1.2), who developed a model of their own, which is simpler and more testable than some earlier versions.

Thresholds have been found to be much more complicated than this, in theory, in wind tunnels, and in the field. Even in dry sand (moist sand is discussed shortly), surface conditions, such as roughness, the grain-size mix, and other factors, each produce their own thresholds, and these may vary in time and space

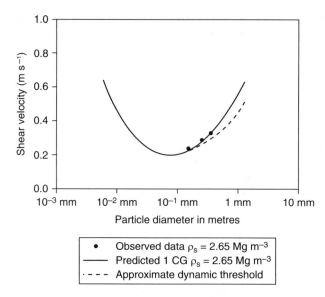

Figure 1.2 Threshold curve for the start of motion (the static threshold) of particles of the density of quartz (observed and modelled data) (Cornelis and Gabriels, 2004). The curve for the dynamic threshold is added and very approximate. Reprinted with permission from John Wiley & Sons.

(Baas, 2007). The following list, therefore, is far from exhaustive. Different thresholds occur when: (1) saltation reaches the intensity at which it can move sand in reptation (shortly); (2) bouncing grains are powerful enough to disperse clods, pellets, and crusts, of varying cohesiveness (Hu ChunXiang *et al.*, 2002) (also shortly); (3) ripples appear (Chapter 2); and (4) ripples move from a 'subcritical' shape (with gentle lee slopes), to a 'supercritical' shape, in which there are small slip faces (Hoyle and Woods, 1997).

Even within one of these groups of threshold, there is a range of behaviour. In a wind tunnel, the static threshold occurs at a spread of velocities, from that at which particles begin to rock back and forth; start rolling; or are lifted from the bed; or, at a larger scale, between the point at which there are only a few flurries of movement in response to bursting turbulence and the point at which the whole bed is mobile. If the wind is accelerated in the wake of even quite small roughness elements, local entrainment occurs at velocities lower than the ambient threshold (Sutton and McKenna Neuman, 2008). The difference between early, sporadic movement and the movement of the whole bed produces static thresholds with a wide spectrum of values. It would be better to choose a probability density of values, although this is seldom done (Williams *et al.*, 1994).

In the field, thresholds are yet more elusive. Measurement is more complex (both of the wind and of blowing sand): winds are gustier; sediments have more sizes and densities; controls like moisture or crusting can limit the supply of loose particles (later); there is far greater spatial variability in all these controls; and features like sand streamers (shortly) complicate measurements of sediment movement. In one field experiment, sand was seen to move at a velocity below a theoretical threshold (calculated from grain size and wind speed), almost certainly because of high instantaneous wind velocities (Rasmussen and Sørensen, 1999). In another experiment, now on a wet beach, it was found that each size and size-mixture of sand, and each set of environmental variables, had its own threshold (Wiggs et al., 2004a). Many of the problems involved in measuring thresholds in the field may be overcome by using terrestrial laser scanning, which can quickly and non-invasively measure surface topography, moisture content of the surface, and perhaps even the sizes of grains in saltation (shortly; Nield and Wiggs, 2011). A recent study, based on measurements in the field, discovered yet another cause of variability: the way in which the threshold is calculated from data (Barchyn and Hugenholtz, 2011).

Grain size

This section continues, but does not, even yet, complete, the explanation for why most dunes are sandy. The relationship between the static threshold and the grain size of sand is shown by a simple experiment: trays, containing grains of a succession of sizes of particle, each with only one size, are exposed, one by one, to increasing wind speeds in a wind tunnel, and the threshold velocity (u_t) or threshold shear velocity (u_{*_t}) at which each size of particle begins to move is plotted against its size. The results of such an experiment are shown in Figure 1.2.

The most important (and most obvious) characteristic of the curve in Figure 1.2 persists, whatever the definition of the static threshold and by whatever means it is verified or theorised: there is a minimum value of the velocity at a grain size of ~0.07 to ~0.1 mm (somewhere between medium and fine sand on the Wentworth scale, earlier). Both values are different in higher or lower pressures and temperatures, as at altitude, near the poles, or on Mars, Venus or Titan (Figure 12.1).

The increase in the threshold on the right-hand side of Figure 1.2 is what one might expect intuitively, and by the earlier explanation for the effect of gravity: bigger, heavier grains need stronger winds to move them than do smaller, lighter ones. But, even for coarse grains, there are complications, the main one being when there is a mixture of sizes. A wind-tunnel study found that the greater the spread of particle sizes in a mixture, the greater was the difference between the static and dynamic threshold (Nickling, 1988). Coarse particles are usually the first to move from a mixture with finer grains, probably because they stick up further into the wind (Martz and Li, 1997). The finer grains in a mixture have a higher threshold of movement than in a one-size sandy bed, because of the sheltering effect of the larger grains (Komar, 1987, RL).

The increase in the threshold on the left-hand side of Figure 1.2 is, of course, because of the increasing effects of the cohesive forces on finer grains (earlier). In short: the finer the particles, the faster the wind must be to move them. This is true in the conditions of the experiment, and in many situations in nature, but does not hold where a bed of fine particles is being bombarded by saltating sand, in which case clods and crusts (shortly) are broken down to dust and raised by the wind.

The slope of the bed

This has a small effect on the threshold values for most sands and in most field situations. If the slope is ≥15°, a slope that is seldom exceeded on the windward slopes of desert dunes (perhaps for this reason, Chapter 3), the threshold is not materially affected. There is, however, a distinct slope effect with coarse sands: field measurements (as reported by Bagnold, 1941, p. 220) and modelling have found that particles coarser than 0.23 mm in diameter (fine sand) are unable to climb a 20° slope normal to the wind at common wind speeds (Tsoar et al., 1996). Although this slope is steeper than many of the windward slopes of dunes, and the sand is not as coarse as many, this may explain why there are accumulations of coarse sand at the base of many dunes. A study of grain-size variations along transects over dunes in the Taklamkan found very distinctly coarser sands on the lower slopes (Wang Xunming et al., 2002c).

The dynamics of water content

It must be true (given earlier arguments) that the amount of water held in a sediment has a strong effect on cohesion, and therefore increases the threshold of movement, but it is difficult, perhaps impossible, to measure the full effect. This is because moisture content in the field and even in the wind tunnel is not constant beneath a surface subject to a wind strong enough to raise particles, because the winds also take moisture.

The various models of the interactions of moisture and the speed of the wind have been reviewed by Cornelis and Gabriels (2003b; Figure 1.3), who developed their own model and tested it in a wind tunnel. This curve would probably be different for different grain sizes of sand. For example, the coarse pores in a coarse sand would lose moisture more quickly, and the capillary replacement of water from lower in the sediment would be slower.

Research in wind tunnels delivers far smaller water contents at the threshold than do field experiments. In one field experiment on a beach of well-sorted coarse sand, the threshold lay between 4% and 6%, which is much greater than has been found in experiments in wind tunnels for sand of the same size (Wiggs et al., 2004a). The reasons are not hard to find. In the field, there is much greater variation in water content, caused by varying inputs of rainwater or dew, and greater variation in output by upward movement (by capillarity) from groundwater; and varying output by evaporation, itself accelerated by the wind, insolation,

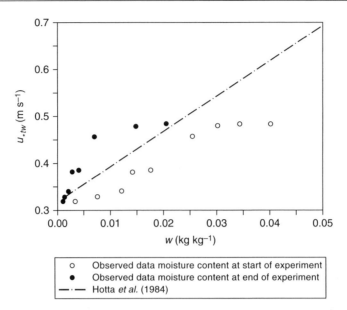

Figure 1.3   Static threshold for grain motion (u_{*tw}) related to the gravimetric water content of the sediment at the start and end of an experiment. Only the results predicted by the model of Hotta *et al.* (1985) (as the best fit of many models) have been retained (Cornelis and Gabriels, 2003). Reprinted with permission from John Wiley & Sons.

high temperatures, drainage under the influence of gravity and capillary rise. The threshold over wet sands varies much more rapidly than that over finer wet sediments, because the higher porosity of sands allows them both to imbibe water more quickly and, if there is not a shallow water table, to drain more quickly. Sands on the surface also dry out more quickly because capillarity in the large pores between sand grains is too weak to bring in water laterally or from below to replace water lost to evaporation. Thresholds may be passed suddenly if water is suddenly lost from the equally sized pores between well-sorted sand. In some circumstances a wet surface can even be hard enough for saltons to bounce off it, thus increasing the intensity of saltation (shortly; McKenna-Neuman and Maljaars Scott 1998).

For these reasons, the threshold in sands may have a daily cycle, according to variations in temperature (Stout, 2003). On a wind-eroding beach, moisture content may also vary with the state of the tide, sea spray and the content of hygroscopic salts. In a set of field experiments on a beach, the threshold was closely correlated with the moisture content at the upwind end of the beach but not at sites further downwind, where bouncing sand drove entrainment (Davidson-Arnott *et al.*, 2008). Erosion and transport create their own spatial and temporal patterns on moist surfaces. If the removal of a layer of dry sediment exposes wet sand, entrainment is suddenly checked, and does not pick up again until the new surface dries. The result is the pulsing of erosion (Pease *et al.*, 2002).

Complexity increases at extremes. At very low temperatures, sand may be released only if the rate of sublimation of ice is sufficient, and this rate itself depends on a number of factors including the temperature, local wind speed, air humidity and ice content of the sediment (Speirs et al., 2008a). In cold wind-blown environments, the patchy growth or thawing of needle ice may withhold or release sediment to the wind, as in parts of South Island, New Zealand, and on some Chinese dry lakes (McGowan, 1997; Mu GuiJin et al., 2002). Some simple statements, notwithstanding, can be made with a modicum of confidence: there is no particle movement at soil water contents over 75%; and the inhibiting effect of moisture is less in coarse than in fine sand (Cornelis et al., 2004b; Ishizuka et al., 2005).

Crusts

Crusts (~2–3 mm thick) are very common on dunes, even in some very dry climates, and can seriously hamper entrainment. There are two types. Abiotic crusts are created by the impact of raindrops, or after evaporation (Ishizuka et al., 2008). Both these processes can produce crusts, even on a bed of glass beads. There are stronger abiotic crusts in the presence of salts, such as gypsum, and where dust has been added to the surface (Scheidt et al., 2010). Abiotic crusts offer less resistance to bombardment than do most biotic crusts (McKenna-Neuman and Sanderson, 2008).

Biotic crusts are formed by animalcules, mosses, lichens, liverworts, fungi and cyanobacteria, whose organic filaments or glues bind the sand (Malam-Issa et al., 2001). On beaches, organisms may aggregate patches of sand, according to subtle ecological differences in moisture and disturbance (Maxwell and McKenna-Neuman, 1994). In the dry, seasonally very cold, Gurbantunggut Desert in far north-western China, where the mean annual rainfall (MAR) is only 70–150 mm, biotic crusts cover 30.5% of the crests of dunes, 48.5% of the upper slopes, 55.5% of middle and lower slopes and 75.5% of interdune areas (Chen YaNing et al., 2007). On dunes in the Negev in Israel (MAR 95 mm, and a much milder winter), morning dew is sufficient to sustain an algal crust, which offers very effective protection against entrainment (Kidron et al., 2009).

The strength of biotic crusts varies from species to species (Godinho and Bhosle, 2009). They are both thicker and more flexible than abiotic crusts and seldom crack or fall apart. If cracks do appear, their frayed edges are the most vulnerable to entrainment by bombardment (Langston and McKenna-Neuman, 2005). Biotic crusts develop at a rate dependent on the environment and the crust-forming species. The rate of development accelerates only after a few years, and some crusts do not mature for ~20 years and can therefore fully develop on already very stable (and undisturbed) sites (Belnap et al., 2009); but disturbed and therefore rougher crusts may and raise the threshold of entrainment.

Pellets

Pellets are aggregates of sand size. Windblown pellets have been reported (or inferred) from: Devonian deposits in Scotland; Late Pleistocene sand dunes in Indiana; Late Pleistocene loess; agricultural fields; Mars (Chapter 12); and many other situations (Rogers and Astin, 1991; Mason *et al.*, 2003a; Kilibarda *et al.*, 2008; Colazo and Buschiazzo, 2010). Their binding agents include micro-organisms, clays, salts and electrostatic forces. In sufficient concentration, sand-sized pellets can be built into dunes (Chapter 4), but most aggregates never reach dunes because they break down quickly in transport and are then dispersed as dust. The abrasion of pellets is faster where most of the saltation load is unaggregated sand (Hagen, 1984). In Western Australia, much stronger pellets, known as 'spherites', have been reported from near-coastal dune sands (Killigrew and Glassford, 1976). They are held together by various cements. The most robust survive in many environments.

Sand in Motion

Once raised, particles travel in four ways, which, in order of increasing velocity, are creep (and related near-surface activity), reptation, saltation and suspension (Figure 1.4). Reptating particles are reptons; saltating ones, saltons; suspended ones, dust. This account begins with saltation because it drives all the others.

Saltation

Saltation is the leaping of wind-driven grains. In the controlled environment of a wind tunnel, and with rounded sand, most saltons are ejected at an angle between 50° and 80° (forwards) from the surface (Li Wanqing and Zhou Youhe, 2007;

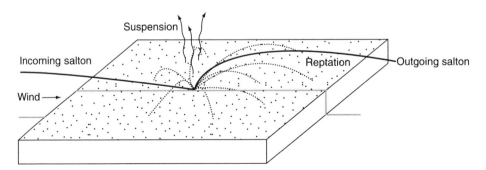

Figure 1.4   Modes of grain transport in the wind (partly based on data from high-speed filming by Mitha *et al.*, 1986). The terms are explained in the text.

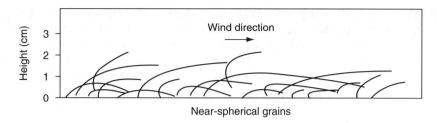

Figure 1.5   Trajectories of near-spherical saltons (Willetts, 1983). Reprinted with permission from John Wiley & Sons.

Figure 1.5). The modal lift-off angle is ~20° to ~40°. No take-off angles less than 20° have been observed. Take-off angles are greater with less rounded grains (Willetts and Rice, 1985). Ejection occurs at a higher velocity where saltons meet an upward-sloping surface, as on the windward side of a ripple or a dune (Willetts and Rice, 1989). A few saltons leap back into the wind, most of them having rebounded from the upwind side of a protrusion; there are more backward leaps at higher wind speeds (Dong Zhibao *et al.*, 2002d). The time of flight is ~50 ms, and a hop on Earth is ~50 cm long (much greater on Mars; Chapter 12). Saltons spend a large part of their flight near the top of their trajectory, which explains the everyday observation that saltating grains seem to 'float' above the surface of a beach or a sandy desert. Angles of return for common sizes of sand are between 9° and 15° to a flat surface (Rice *et al.*, 1996a).

In wind-tunnel experiments, launch velocity does not alter much with u_*, which may be because increasing u_* drives more grains in saltation, rather than increasing the velocity of individuals (Namikas, 2003; Rasmussen and Sørensen, 2008). When they reach the highest point of their trajectory (~80 mm above the bed), the velocity of 0.242 mm and 0.320 mm diameter saltons reaches that of the wind. As they descend through the slower wind nearer the bed, the ratio of their velocity to that of the wind around them steadily increases, until it reaches ~2 when they meet the bed, all irrespective of grain size. Because of the added impetus given by collisions, the forward speed may exceed the wind speed. The denser the cloud and the faster the wind, the more collisions there are, and these widen the distribution of the velocities of the grains. Finer particles have less chance of collision because they travel high in the wind, where there are fewer grains in motion (Dong Zhibao *et al.*, 2005). Mixtures of sizes of sands have little effect on the characteristics of saltation, such as ejection speeds, ejection angles and the mass flux profile (Xing Mao *et al.*, 2011).

In general, smaller particles travel faster, but a few large saltons leap high, perhaps because of their greater momentum, and because their lower specific surface area suffers proportionally less drag (Jensen and Sørensen, 1986). In a sand storm in which fine particles of the order of 0.1 mm rose only a few centimetres, a few with diameters of 1.0 mm reached 1.5–2 m above the surface (Sharp, 1964). In a severe windstorm in the San Joaquin Valley in California, in which wind speeds at

10 m above the surface reached 53 m s^{-1}, particles of 23 mm diameter were imbedded in a telegraph pole 0.8 m above ground (Sakamoto-Arnold, 1981).

Almost all observations of saltons have been in wind-tunnel experiments. The few that have been made in the field show more complex behaviour, for several reasons, principally the variations in roughness, and the hardness of the surface and the coverage of crusts (Stout and Zobeck, 1996c).

Streamers and other medium-scale patterns of saltating sand

Streamers of bouncing sand, so familiar on windy beaches and dunes, may be related not to the burst–sweep process but to the downward propagation of larger, higher turbulent structures. In one experiment, streamers on a fairly uniformly dry sandy surface were ~2 m wide and spaced at ~1 m. These dimensions appear to be independent of the velocity of the wind and perhaps also of the nature of the bed. Streamers intertwine and bifurcate, in ways that are more complex in higher winds. They may have a dominant role in initiating and maintaining the movement of sand, implying that there is more yet to learn about sediment transport by the wind, but their study presents some major challenges, and they have been little researched (Baas, 2008).

There are other horizontal patterns of transport on Sahelian fields, and doubtless in many other similar situations. They are created by roughness elements such as bushes (Visser et al., 2004b). Another pattern, on a dry lake in western Queensland, is also likely to be common. It was generated by variations in the availability of sediment. The horizontal variability of supply was greater at low wind speeds than at high speeds, when many sources yielded sediment (Chappell et al., 2003c).

Reptation

Reptation (or 'impact creep') is the 'splashing' or low hopping of grains that have been dislodged by descending saltons (Figure 1.4). Reptons hop just once. They have much less momentum than saltons, so that, when they return to the surface, they neither rebound nor disturb others, although they may roll a few millimetres. The size distributions of reptons and saltons have substantial overlap, and grains continually pass between the two modes of travel, but the velocity distribution of reptons is heavily skewed towards small velocities, with a long tail of faster ones, whereas that for saltons follows a peaked Gaussian distribution (Anderson, 1987b). The velocity of reptons is only weakly dependent on u_* but strongly related to their size (Schwämmle and Herrmann, 2005a).

Reptons absorb more of the energy of the saltons that disturb them than is taken by outgoing saltons, and at any one time most of the grains in motion are reptons (Anderson *et al.*, 1991; Rice *et al.*, 1995). The transport rate in reptation, being powered by saltation, scales with u_*, but its contribution to the overall transport rate declines as the overall transport rate increases. Because they are not driven directly by the wind, reptons respond to the effect of slope, where saltons cannot. Chapter 4 shows how this behaviour may play a major role in controlling the shape of transverse dunes.

Creep

Before the discovery of reptation (as it is now understood), the term 'creep' was applied to all near-surface movement (as by Bagnold, 1941, pp. 33–35). 'Creep' is now reserved for two types of slow surface movement, which, like reptation, are caused wholly by bombardment (rather than wind shear). The first is the rolling of coarse particles driven by the impact of saltons (unlike the small leaps of reptating grains); the second is rolling under gravity, as into craters created by the saltation impacts, or down the lee sides of ripples (Chapter 2). Most particles in creep are coarser than those in saltation or reptation, but the size of creeping grains is also a function of the prevailing mix of grain sizes and of u_*. Data from fast-shutter images show that with u_* at $0.48\,\mathrm{m\,s^{-1}}$, 0.355–0.6-mm-diameter sand grains creep at ~$0.005\,\mathrm{m\,s^{-1}}$. Grains of the same size begin their journey together but rapidly disperse, as some move faster than others; in one experiment, dispersal was complete within 3 min (Willetts and Rice, 1985b). Some creeping grains (like some reptons and saltons) are buried for long periods of time, many in ripples (Barndorff-Nielsen *et al.*, 1982). Partly because of differences in sampling methods, partly because of the use of different sizes of sand and of different values of u_* in experiments, and partly because of the occasional lumping of reptation with creep, estimates of creep as a proportion of the total wind load have varied from 6.5 to 50% (Wang Zhenting and Zheng Xiaojing, 2004). In Wang and Zheng's model, the proportion of the load in creep is a high proportion of the total flux rate at low u_* but rapidly declines at higher u_*.

Other near-surface activity

A closer look finds that yet more activity occurs near the surface. The first is the tunnelling beneath the surface of high-velocity saltons on their return (Willetts and Rice, 1985b). The second is a compressional-dilational wave that radiates from the point of impact of a salton, shaking the sand to a depth of about five grain diameters. Models show that the shaking raises large particles by rotating and ratcheting them against smaller ones, which lifts rougher particles more quickly. All this activity may explain the very thin layer of exceptionally well-sorted coarse sand on an erosional surface (Sarre and Chancey, 1990). Third, bombardment may elevate some grains to positions where they are more vulnerable to dislodgement (Iversen *et al.*, 1987). Countering the destabilising processes, bombardment sometimes consolidates and partly immobilises the surface.

Suspension

Suspension depends on both u_* and the fall velocity of a particle, u_f, which is a function of the balance between the weight of the particle and the drag of the air upon it (Stoke's Law). The vertical velocity in turbulence near the ground is directly related to u_*, so that, as a rule of thumb, if $u_f < u_*$, particles stay aloft. Observations at quite a spread of values of u_* show a sharp transition to suspension or an intermediate behaviour, 'modified saltation', where grain size is less than ~0.1 mm (Nalpanis, 1985). There is therefore a fairly clear and common distinction between the behaviour of particles of silt and clay size (or dust), which can be held aloft at many common wind speeds, and that of sands, which rarely go into suspension (except in the lee of dunes, Chapter 3). This completes the explanation for why dunes are sandy.

The vertical distribution of load and grain size

Measurements of the movement of a mixture of sizes of sand using a Phase Doppler system have now shown that the maximum flux is just above the surface (Figure 1.6); the peak height increases with, and is more marked in, stronger winds. Above this near-bed convexity, the profile adopts a form of curve more like Bagnold's (1941, p. 63; Dong Zhibao *et al.*, 2006a), which showed a near-bed plateau below a smooth curve. The near-surface peak is probably the mean height of reptating grains (Ni JinRen *et al.*, 2003a).

The top of the saltation 'cloud' proper (in which there are many more particles in motion) is between 14 and 15 cm above the surface. In most cases, with the exceptions noted earlier, the modal size of grains in transport declines smoothly with height above the bed. Different values of u_*, different grain sizes and grain-size mixtures, and other variables affect the vertical distribution (Dong Zhibao and Qian GuangQiang, 2007b).

The saturation length

As well as responding to the speed of the wind, the quantity of sand carried responds to a change in the character of the surface over which it blows. The response to a change from a hard, compact surface to a loose sandy surface is of special interest because it probably determines the minimum size of a dune, and may be one of the main determinants of the shape of a mature dune (Chapter 3). The adjustment to a loose surface is the outcome of changes in many different

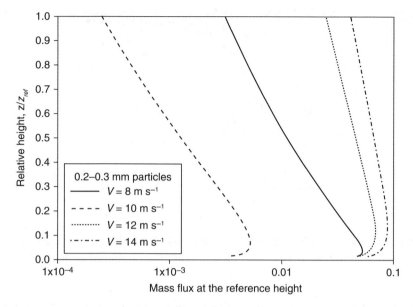

Figure 1.6   Vertical distribution of dimensionless sand flux, for 0.2–0.3 mm particles, at different free-stream velocities (Dong Zhibao *et al.*, 2006a, which contains data for many more particle sizes).

processes, such as: the length needed to complete the release of new grains into the wind; their acceleration to a new steady velocity; the change in hop length of saltons on the new, loose surface; and the decelerating effect on the wind of the increase in the load of sand it is carrying. It is the slowest of these processes that determines the ultimate distance to overall adjustment.

Bagnold (1941, pp. 180–183) discovered what is now known as the saturation length in his wind-tunnel experiments. More recently, the phenomenon has been studied, also in a wind tunnel, by Andreotti and colleagues (2010). As shown on Figure 1.7, they found two phases in the overall response: first, an exponential increase in the load as new grains are released, which they labelled '$L_{1/4}$'. $L_{1/4}$ becomes rapidly shorter at higher wind speeds and is negligible in high winds; second, a stage of steadily increasing load towards a new plateau, over a distance that they labelled 'L_{sat}' or the 'saturation length'. In their experiment, L_{sat} was ~1.5 m long. This is well short of Bagnold's result, a difference they attribute to Bagnold's system for measuring sand flux with spring balances, compared with the spatial resolution of 10 cm for their own measurements of sand flux. L_{sat} is not sensitive either to changes in wind speed or to whether the approaching wind does or does not carry sand. Andreotti and colleagues noted that:

$$L_{sat} \approx 4.4 \rho_s / \rho_f d,$$

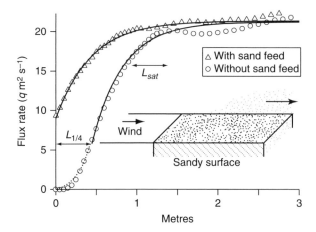

Figure 1.7   Saturation length or L_{sat} (the distance taken for the load carried by the wind to adjust it passes from a hard, cohesive to a sandy surface). The terms $L_{1/4}$ and L_{sat} are explained in the text (redrawn from Andreotti *et al.*, 2010).

where ρ_s and ρ_f are the densities of the sand and of the air, and d is the sand grain diameter

As can be seen on Figure 1.7, Andreotti and colleagues' experiment revealed a minor fluctuation immediately after the first peak in the curve of L_{sat}. Bagnold's finding (1941, p. 182), which was more or less confirmed by Spies and McEwan (2000) and Arnold (2002), was that there were more distinct and more fluctuations beyond a first peak, which continued for ~7 m downwind of the upwind edge of the patch of sand. The explanations for the fluctuations offered by McEwan and Arnold were: (1) the equilibrium between the cloud of saltating sand and the wind above it takes time to propagate up through the saltating curtain (Arnold, 2002); (2) saltation trajectories shorten as the number of grains in the wind increases, and this brings down the transport rate (Almeida *et al.*, 2007), which could be because; (3) the buildup in the rate of collisions between saltating particles, which disperses the available energy, as shown later by Dong Zhibao and colleagues (2005).

The fetch effect

This is a longer spatial and temporal adjustment between the wind and its load. The adjustment occurs on surfaces, such as agricultural fields, dry-lake beds or tidal beaches, where the sediment on the surface contains a significant content of silt and clay, or is of variable wetness, both being situations where the surface releases sand slowly and which therefore delays the point of maximum load, which is not achieved until the 'fetch' distance (Delgado-Fernandez, 2010). There is no fetch effect where there is an adequate supply of sediment, as on a dry beach, or in most desert conditions.

The response of a loose bed to erosion by the wind

Because the fine particles on a bed of unaggregated particles are removed before the coarse ones, the surface of the eroding bed coarsens (Bagnold and Barndorff-Nielsen, 1980), which raises the threshold of movement and reduces the erodibility of the bed. The same is true where an eroding surface is covered with loosely cemented aggregates or a crust. In that case also, the fine particles are the first to be released, at least until the crust or pellets have disintegrated, although this point is seldom reached within a single wind storm (Stout and Zobeck, 1996).

The Transport Rate

The transport rate is commonly denoted q (or Q), defined as the mass of sediment passing through a plane perpendicular to the wind, of unit width and of infinite height, per unit time (in kg (m-width)$^{-1}$s^{-1}; or m^3 (m-width)$^{-1}$s^{-1}).

Because of the difficulties in measuring the rate of blowing sand, effort has been focused on the prediction of the transport rate from wind data (which are easier to collect, and are often collected routinely for other purposes). For all the effort, however, there is still no universally accepted relationship. This is partly because of the evolution of ideas about the importance of u_* (earlier), but some pessimists believe that the transport rate is inherently indeterminate, given the number of variables involved (Bauer et al., 1996; Smith and Stutz, 1997). Indeterminacy can be kept to a minimum in a wind tunnel and is manageable where a surface in the field is dry and level, and where there is a copious supply of well-sorted quartz sand of appropriate size, but these requirements are rarely met. In the field, it is common to find variations in: grain size, hardness of the surface (shortly), slope and curvature of slope, microtopography, wetness, crusting and other controls on the availability of sand.

The modelling of the relationship between the speed of the wind and its load has been approached in four ways (Ni JinRen et al., 2004): those based on (1) relations between the paths of saltating grains and the wind (Bagnold's approach, 1941, pp. 64–71); (2) the relations between the concentration of grains and their trajectories (Owen's approach, earlier); (3) linkages between trajectories with grain-surface collisions and the corresponding adaptation of the wind; finally, there has been (4) wholly empirical curve-fitting to experimental observations from either wind-tunnel or field measurements (for example, Liu Xianwan et al., 2006). All of these approaches have relied heavily on data from experiments in wind tunnels.

Most of the models relate the transport rate (q) and shear velocity (u_*) in the general form:

$$q = \alpha\left(u_* - u_{*_t}\right)^b ,$$

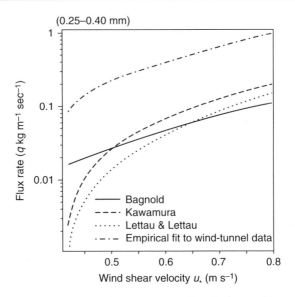

Figure 1.8 Comparison of the performance of selected transport formulae for particles 0.25–0.40 mm (medium sand) in size (Liu XianWan *et al.*, 2006).

where the constant $\alpha = \sim 2$; u_* is shear velocity; u_{*_t} is threshold shear velocity; and b is an exponent (Almeida *et al.*, 2007).

Comparisons of the predictions of various models show that they diverge in their predictions, some wildly. Figure 1.8 shows divergence of the relationship of wind speed and sand transport for one grain size (curves for many more grain sizes are given in Liu Xianwan *et al.*, 2006). The divergences are wider where the experiments have modelled extreme conditions, as for very fine and very coarse sands, or at very high or very low wind speeds. A field test on a windy Irish beach found most of the models gave poor results, probably because they could not account for surface moisture. Bagnold's (1941, p. 67) and Zingg's (1953c) models, both aging, were the best of a bad bunch (Sherman *et al.*, 1998).

The model that has been most widely used in studies of dunes (and especially in calculating the directional variability of sand movement, Chapter 4), is Lettau and Lettau's (1978):

$$q = C\left(\rho_a/g\right)u_*^3\left(1 - u_t / u_z\right)t^{-1},$$

where q is the discharge rate of sand in grams (m-width)$^{-1}$; C is an empirical constant related to grain size, commonly ~ 6.5; ρ_a is the density of the air; u_t is the impact threshold velocity; u_z is wind velocity at height z; and t is a specified time period.

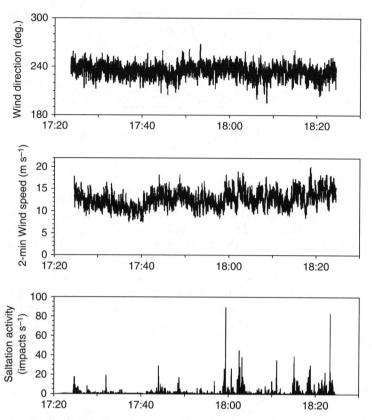

Figure 1.9 Variation in the speed and direction of the wind in the field resulting in a very variable rate of transport, labelled 'intermittent saltation' (Stout and Zobeck, 1997). Reprinted with permission from John Wiley & Sons.

New transport models continue to be proposed. One of the latest is designed to be used for small fluxes (Almeida *et al.*, 2006).

At the small scale, the transport rate has been shown to be remarkably responsive to changes in wind speed. If the frequency of variation in velocity is ~1 Hz, the transport rate tracks the speed of the wind, but it does not follow higher frequency variations (Butterfield, 1999a). In the field, wind speeds and directions usually fluctuate wildly so that sand may be moving only part of the time. In one field experiment, saltation was active for only 26% of the time (Stout and Zobeck, 1997; Figure 1.9).

Shapes, densities and mixtures of size

At low u_*, platy grains (as of shell sands) have higher transport rates than more rounded ones, but at low u_*, platy grains have markedly lower transport rates, size, density and sorting being held constant (Willetts *et al.*, 1982) (Figure 1.10).

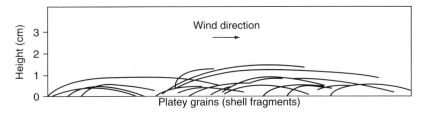

Figure 1.10 Trajectories of platy saltons (Willetts *et al.*, 1982). Reprinted with permission from John Wiley & Sons.

There are many more dimensions to the shape of particles, each having an effect on transport in the wind. When shapes are complex, as are the shapes of the minute bivalves of the genus *Mya*, the wind, unaccountably, sorts left- from right-curving shells (Cadée, 1992). Denser sands (as in those composed of magnetite) have a markedly lower transport rate at low u_* and a somewhat lower rate at high u_* (size, sorting and shape being held constant; Williams, 1964). In some size mixtures, large saltating grains dislodge smaller ones, in which case, the transport rate of the coarse fraction is higher than that of the mean for the mixed size (Iversen and Rasmussen, 1999b). The long-term development of roundness in windblown sands is discussed in Chapter 9.

Hard surfaces

Bagnold (1941, p. 71) observed that saltation trajectories are higher and longer over hard surfaces, as of rock, or over a surface strewn with pebbles ('desert pavement'), than on sandy surfaces. Hardness is measured by the coefficient of restitution, being the ratio of the velocity of an object before it hits the surface to its velocity on rebound. At the same wind speed, and given a sufficient supply of sand, therefore, more sand is in movement over hard surfaces than over loose sand. There is also a more distinct peak height of flux over the harder surfaces, which increases in height downwind of the windward edge of a patch of gravel (Dong Zhibao and Qian Guangqiang, 2007b).

Rough surfaces

The transport rate is also dependent on the roughness of a surface, as shown in a field experiment at a contrast in roughness between an alluvial plain and a very rough lava field in the Mojave Desert by Greeley and Iversen (1987). When the wind first crossed from the alluvium to the lava field, there was a surge in stress (and hence sand-carrying capacity), after which the stress settled down, but to a higher level than over the smooth alluvium. There was another adjustment where the wind passed back from the rough to the smooth surface. Here, too, the change was at first abrupt, both in shear stress and in transport capacity, but stress and carrying capacity again slowly picked up. The transport rate is further enhanced by the greater air turbulence over the rough surface.

Moisture, temperature and humidity

For all the inherent, some perhaps insuperable, problems of studying the details of the effect of moisture on thresholds (earlier), there have been some constructive studies of its effect on transport rates. Many have shown, for example, that when shear velocities are well above the threshold on a dry surface, the moistness of the surface has little effect on the transport rate. This is partly because of the increase in the drying capacity of the wind. On a Dutch beach, sand flow did not reach near the rate predicted by transport equations from wind-speed measurements, until relative humidity fell below 85% (Arens, 1996b). Moisture is thought to be the main reason that few measurements of sand flux over a wet beach agree with the predictions of mathematical models (Bauer *et al.*, 2009).

Rain

Intense wind-driven rain increases the transport rate by splashing up particles into the path of the wind and by lengthening saltation trajectories (by, on average, three times) (Erpul *et al.*, 2004b). The effect is strongest on saturated surfaces, as on beaches exposed by a fall in the tide, or wet, bare upland peats (Foulds and Warburton, 2007a). In these cases, driving rain substantially enhances sand transport. The effectiveness of the rain-driven process depends on the size of the raindrops, the slope of the surface, the angle at which the raindrops meet the surface (raindrops driven by a wind are more effective than vertically falling rain; Erpul *et al.*, 2005) and the grain size of the sand (Furbish *et al.*, 2007).

References

Komar, P.D. (1987) 'Selective grain entrainment by a current from a bed of mixed sizes: a reanalysis', *Journal of Sedimentary Petrology* 57 (2): 203–211.
Wentworth, C.K. (1922) 'A scale of grade and class terms for clastic sediments', *Journal of Geology* 30 (3): 377–392.

Chapter Two
Ripples

Unlike ripples beneath other fluids and in other materials (under flowing water, under oscillating waves, on ice surfaces and so on), and also unlike dunes, the dominant process in the formation of windblown ripples is reptation (Chapter 1). Windblown ripples cover all dry, bare, loose sandy surfaces (Figure 2.1), with two exceptions, both minor: where the sand is too coarse to reptate in prevailing wind speeds; and where there is grain-fall into zones with gentle winds (as to the lee of dunes).

A windblown ripple starts to form when a forceful salton blasts a crater as in a loose sandy surface. The upwind-facing slope of the new crater then captures more of the incoming saltons than did the preceding level surface, and these drive more reptons (Chapter 1). The new reptons then contribute to the growth of a mound downwind of the crater, whose growth is self-perpetuating (more saltons captured, more reptation), until the lee slope of the new ripple is steepened beyond the angle of repose of the sand. Sand then tumbles down the steep slope, and the ripple moves forward, maintaining its form, much as do dunes (Anderson and Bunas, 1993; Chapter 3; Figure 2.2). The introduction of a 'slip-face' mechanism to a model of ripple formation improves its resemblance to real ripples (Caps and Vandewalle, 2001).

Growth continues for about 10 min until the ripple achieves the first of a sequence of 'equilibrium' forms (Seppälä and Lindé, 1978). The spacing between a succession of these 'proto-ripples' then slowly increases, partly because smaller ripples collide and merge with larger ones, as with dunes (Andreotti *et al.*, 2006; Chapter 3). The rate of growth of wavelength then quickly declines towards a second 'equilibrium', whose controls are poorly understood, although the process of growth in wavelength has been observed in many models (Manukyan and

Dunes: Dynamics, Morphology, History, First Edition. Andrew Warren.
© 2013 John Wiley & Sons, Ltd. Published 2013 by John Wiley & Sons, Ltd.

Figure 2.1   Selection from the large range of ripple wavelengths and patterns. (a) The most common pattern of ripples in well-sorted sand on the windward slope of a dune. (b) An unusually widely spaced, and unusually straight ripple pattern in somewhat coarser sand. (c) Ripples in sand with moderately mixed sizes, and with many 'defects' (explained in the text). (d) Mega-ripples in coarse sand at the base of a dune, juxtaposed with shorter-wavelength ripples in the fine sand on a dune. Image (b) is reprinted with permission from the Royal Geographical Society (with the Institute of British Geographers).

Prigozhin, 2009). Wavelength is greater on windward slopes and least on the downwind slopes of dunes (Werner *et al.*, 1986). Mega-ripples (shortly) take longer to reach a larger equilibrium size and spacing (Yizhaq, 2005).

The angle of the windward slope of a ripple is limited to about 20°, perhaps because above that angle, fewer incoming saltons and the reptons they disturb would rebound downwind (Rumpel, 1985). The windward slope is gently convex, with a convexity that is relatively broader than on dunes (Werner *et al.*,

Wind

(a)

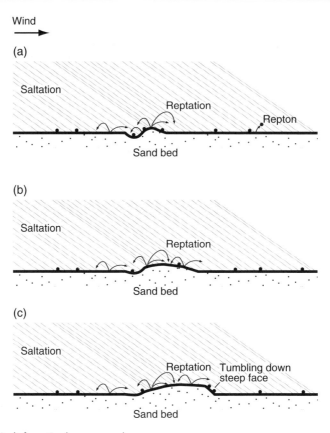

Figure 2.2   Ripple formation (many sources).

1986). Some ripples have sharp crests and/or brinks (these terms are explained in Chapter 3 in reference to dunes). The leeward slope may reach ~30°, as on dunes.

The celerity of ripples (celerity = speed of movement of ripples or dunes, Chapter 3), and the initial and final wavelength are all linearly related to u_* (Andreotti *et al.*, 2006). The celerity of ripples in fine sand with a spacing of ~10 cm, and in a wind of ~9 m s^{-1}, has been measured at ~3 cm min^{-1}. In the same experiment, the movement of nearby mega-ripples was much slower (Lorenz, 2011a; a video of ripple movement is available through that paper). As with dunes, celerity is inversely related to size (Lorenz and Valdez, 2011). As u_* increases, small ripples may cover the larger ones (Seppälä and Lindé, 1978). The wavelengths and heights of ripples are positively related to grain size, and wavelength is related to four times the mean length of saltation (at least in a model; Yizhaq, 2004). Ripples respond quickly to changes in the velocity and direction of the wind (Sharp, 1963).

As the power of bombardment eases towards the crest of a ripple, coarse sand accumulates there (Bagnold, 1941). Sand rolling down the lee slope creates foreset bedding, which then makes up the body of the ripple. Fine material may be shaken down into the body by the forward movement of the ripple (Sharp, 1963). The foreset bedding (Chapter 3) may be retained in large ripples but is often 'wavy', unlike the straight laminae on the slip faces of dunes (Hunter, 1977b). A succession of smaller ripples on an accumulating surface forms thin 'pin-stripe', surface-parallel laminae, in which coarse grains overlie fine (Forrest and Haff, 1992). Together with the pin-stripes of the slip faces of dunes, also formed by disturbance, which also have 'reverse grading' (Chapter 3), these are the surest indicators of an aeolian rather than a subaqueous sedimentary environment (Fryberger and Schenk, 1988).

Ripple-to-ripple downwind distances (wavelengths) are between a few centimetres and ~43 m, and the heights are <0.01 m to ~2.3 m (Milana, 2009). Most ripples are aligned at right angles to the wind, but where the direction of the slope of the surface on which they develop is itself not at a right angle, as on many of the windward surfaces of transverse dunes, reptating sand acquires a downslope and lateral component driven by gravity, and ripples are oblique to the wind (Howard, 1977). The significance of this kind of movement to the shape in dunes is explained in Chapter 3.

Subtypes

The most commonly distinguished subtype is the 'mega-ripple' ('granule ripple', or 'ridge'; Figure 2.1d), on whose crests there is a (usually thin) layer of much coarser sand than on smaller ripples. The coarse grains are moved as saltons in high winds but also as reptons in winds below the entrainment threshold of the large grains (Zimbelman et al., 2009).

The cross-section of most mega-ripples is more symmetrical than that of smaller ripples, perhaps because of their greater persistence, which would allow for modification of slopes by winds from many quarters (Fryberger et al., 1992). Bagnold (1941, pp. 156–157) guessed that the mega-ripples (his 'ridges') that he had found in the Sahara had taken centuries to form, but others have been found on railway cuttings, suggesting a much shorter life (Sakamoto-Arnold, 1981), and the flattening and destruction of mega-ripples have also been observed after strong storms, which probably entrained the coarsest grains as saltons (Yizhaq et al., 2004). Increasing shear over the crests of mega-ripples as they grow upward, probably limits their height (Isenberg et al., 2011). Mega-ripples are much more persistent than ripples in finer sand. Mega-ripples have been claimed as a distinct form of ripple, dependent on a distinctly bimodal mixture of sand, which characterise the lower slopes of dunes (Chapter 1).

Distinctions have also been made between supercritical ripples, which have short upper lee faces at up to 30°, and subcritical ripples that do not (Hoyle and Woods, 1997). Wet surfaces, as on beaches exposed by the tide, or any surface just after rain, develop 'adhesion' and 'rain-impact' ripples (Clifton, 1977; Hunter, 1980b).

Chevrons are rare, and are placed here, with ripples, only by virtue of their low amplitude. They are visible on some satellite imagery but are not easily distinguished on the ground. Their wavelength is between 130 and 1200 m, and is locally variable; their amplitude is only 20–30 cm. They move at up to 500 m yr^{-1}. Ten per cent by weight of the grains on southern Egyptian chevrons are 3.5–4.5 mm granules, in a matrix of coarse sand. The shadowing effect of the granules may cause a darkening of the image on the imagery. Unlike ripples and dunes, the steeper slope is upwind. Downwind of some, there is a finer-sand, lighter chevron. Their origin is a mystery (Maxwell and Haynes, 1989).

Models

This section describes some of the many blind alleys and the fewer open-ended alleys that have been proposed as explanations of ripples.

Flow response

In these models, at their simplest, an eddy develops in the lee of a low obstruction formed by some other process. A ripple develops at the downwind end of the eddy, where the full forward force of the wind again reaches the bed; subsequent ripples grow in the lee of the first (Cornish, 1914). The discovery that burst–sweep sequences (Chapter 1), which are manifestations of turbulence at the scale of a ripple, have dimensions much longer than a ripple wavelength leaves this hypothesis with little empirical foundation. Like the models that propose a one-to-one relationship between the dimensions of atmospheric turbulence and the downwind arrangement of transverse dunes (Chapters 3 and 4), these models now have little traction today as explanations of windblown ripples.

Gravity wave

This model applies the Helmholtz–Kelvin mathematical models of gravity waves at the interface of two fluids of different density (like air and water). At its crudest, when applied to ripples, the two fluids were air and loose sand, but it was soon conceded that loose sand does not behave as a fluid (Högbom, 1923c). In a more sophisticated model, the two fluids are dense air near the ground and less dense air above (Kármán, 1947), although Kármán was hesitant in this proposal. A later proposition was that the interface on which waves developed was that between more-or-less sediment-free air above and the cloud of saltating sand below, the waves being transmitted to the bed through their effect on the downward velocity of saltons (Brugmans, 1983), but this model has been taken no further.

Saltation length

An early example of the saltation-length model was applied to windy Irish beaches. Joly (1904) thought he had found a correspondence between the wavelength of ripples and the lengths of the leaps of saltons. In other words, sand bounced from ripple to ripple. Bagnold's (1941, pp. 149–150) observations in his wind tunnel, which were more rigorous, seemed to confirm the model, although he was cautious about the jump-length hypothesis, as was Kármán (1947) in relation to this, as to his other model (earlier). The saltation-length model seemed to explain the greater wavelength of ripples in coarse sand, because coarse sand would provide a better surface for rebound, thus lengthening saltation paths (Ellwood *et al.*, 1975).

Experimental observations later destroyed the empirical basis of the saltation-length model. Saltation leaps, measured in wind tunnels, are 'an order of magnitude' longer than the ripple wavelength (Willetts and Rice, 1989, p. 719), and saltation trajectories that correspond to the wavelength of ripples do not have enough energy to drive reptation or creep (Anderson and Hallet, 1986), but the model is not dead (Miao Tiande *et al.*, 2001b).

Shadow zone

This model, developed by Sharp (1963), merged field observation and simple theory, and is the first of the early models to have survived in part (for example, in Hoyle and Woods, 1997). Sharp's main contribution was his explanation for the spacing of ripples. As a ripple grows in this model, it shields an increasing area on its lee from bombardment, so that the next ripple cannot develop there and must develop progressively further downwind. For Sharp, this also explained the growth in wavelength after initiation, an observation that was at odds with the path-length model: leap lengths in a constant wind should themselves be constant.

Mathematical

These models were developed in the 1980s. They have considerably improved the understanding of ripples and have provided most of the argument in this chapter.

Mathematical models of ripple-forming processes develop expressions for selected parts of the ripple-forming processes, such as: the incoming energy of saltons; the slopes of the surfaces they land upon; the resulting splashing of reptons; the effects of surface curvature; the effects of the increase in velocity of the wind as it is constricted by the growing ripple; the increase in the area of shadow zone downwind of a growing ripple; the tumbling of grains down lee slopes; and so on. Mathematic expressions for some or all of these processes are then combined into a model, which is then analysed to see what kind of wavelengths appear, how this occurs, at what rate patterns stabilise, and to what extent the other characteristics of the modelled ripples, such as speed of propagation, height/wavelength characteristics and so on, compare with real ones.

Anderson (1987a, 1990) was the pioneer of mathematical modelling. Later models in this family include those of Landry and Werner (1994) and Prigozhin (1999). The models have generated some intriguing predictions, such as that ripples might move upwind, even though upwind movement has not (yet) been observed in the field (Kurtze *et al.*, 2000).

Pattern

The patterns of ripples and of dunes differ in several respects: (1) almost all ripples are transverse; there have been reports of linear, near wind-parallel ripples, but they have not been fully studied in the field, in a wind tunnel or mathematically; (2) small ripples in fine sand are generally much less sinuous in plan-view than are dunes (Figure 2.1d); (3) some 'mega-ripples' in coarse sand, on the other hand, are very sinuous, some even more so than transverse dunes (Figure 2.1d); and (4) there are fewer 'defects' in ripples in fine sand than in dunes (Figure 2.1a). In the field, ripple defects have been seen to move forward at about half the speed of the ripples (Lorenz, 2011). Some of these differences are not hard to explain: ripples, having far less bulk than dunes (which may take thousands of years to eliminate defects), can adjust within minutes to a change in wind direction.

The first cellular automata for windblown bedforms were for both ripple and dune patterns (cellular automata as a whole are explained for dunes in Chapter 4). The most important conclusion from these models is that ripples, like dunes, are self-organising systems, whose pattern can be reproduced with a few rules. Pattern formations in ripples and dunes share some of the same mechanisms, for example the negative relationship between dune size and celerity and slip-face activity (Hoyle and Woods in relation to ripples, 1997; Chapter 4 in respect of dunes; and Chapter 5 in relation to pattern formation in dunes). But there are also differences: successive ripples are almost certainly not repulsed by the circulation bubble as are dunes (Chapter 4).

Chapter Three
The Form and Behaviour of Free Dunes

Definitions

By 'form', I mean: the downwind cross-sectional shape of a transverse dune, in conformity with the two-dimensional spatial frame of Part I of this book. 'Form' is distinct from 'pattern', which is reserved for the three-dimensional shapes of dunes and Part II. 'Free' was probably first used by Dobrowolski (1924) to distinguish mobile dunes from 'forced' dunes, which form around obstacles such as bushes, boulders, buildings and bluff hills (Chapter 5). 'Mobile', in this terminology, applies to the dune, not its surface, which may be mobile on an immobile dune. This chapter and Chapters 4–7 are about dunes that are actively forming.

The features described in this chapter are best developed on transverse dunes, which are those created by winds that blow consistently from one direction throughout an annual cycle (Chapter 4). Nonetheless, the features are shared by dunes in more directionally variable wind regimes, even if their slip faces are usually smaller, and their windward slopes usually steeper, because they are prevented from full development by repeated changes in wind direction (Chapter 4).

Early Stages

Start

On Padre Island off the Texas coast, new free dunes reappear every dry season. In their season, patches of sand gather in slight hollows; at changes from rough to

Dunes: Dynamics, Morphology, History, First Edition. Andrew Warren.
© 2013 John Wiley & Sons, Ltd. Published 2013 by John Wiley & Sons, Ltd.

smooth surfaces; or around obstacles like plants. The patches soon break free of their parent surfaces or obstacles, and some grow (Kocurek *et al.*, 1992).

Minimum size

Bagnold (1941, p. 183) maintained that the minimum size of such a dune depended on the behaviour of a wind when it meets a patch of sand. Figure 1.7 shows that adjustment takes a certain time and covers a certain distance. In Bagnold's explanation, a patch shorter than the distance of adjustment is liable to be eroded by a wind that is not fully saturated with sand. Bagnold's 'distance to adjustment' is now known as the 'saturation length' (although Bagnold's and other definitions and estimates of the same phenomenon are different in detail as explained in Chapter 1).

Bagnold's hypothesis (or at least the principles behind it) has now been tested with observations from a Moroccan beach, which has a very constant wind-directional regime (defined in Figure 4.3) (Andreotti *et al.*, 2002a; Elbelrhiti, 2012). On this beach, dunes began as slip-faceless piles of sand (or 'proto-dunes'; Figure 3.2a), which have been reported from many beaches. The minimum size (λ_{min}) was found to be:

$$\lambda_{min} \sim 12 L_{sat},$$

L_{sat} being the saturation length (as in Chapter 1).

Because L_{sat} is related to the grain size (among other things), this means that the minimum size also depends on grain size. The relationship applies also to subaqueous ripples (not subaqueous dunes, later) and to dunes in the thin atmosphere of Mars (Chapter 12).

After they had moved downwind over a distance of a few hundred metres, which took about five days, Elbelrhiti's proto-dunes were transformed to small barchans or transverse dunes, with slip faces between 60 and 100 cm high. By then, the dunes had grown upward, first by adding sand immediately above the brink, so that the slip face grew; sand was later added to the crest, further back, creating a separation of the crest from the brink; slip faces were then ~0.4 m high. Observations of several of these new dunes showed that there was a critical downwind length after which they survived. At 17 m, this critical length is somewhat longer than Bagnold's (1941, p. 182) proposal. Elbelrhiti found that the relation of one of these dunes to its neighbours was another control on its development and size (explored more fully in Chapters 4, under 'barchans'). The downwind length of these dunes was strongly dependent on u_* at low wind speeds, but became less dependent with increasing u_*, and tended to a constant wavelength in higher winds (Andreotti *et al.*, 2010).

A third category of small dune (in addition to proto-dunes and small free dunes) are dunes developed on top of larger dunes by winds from a new direction.

In Morocco, a common wavelength of these dunes atop medium-sized barchans is ~28 m, which compares favourably with a prediction of 20 m in a model. The wavelength of the biggest barchans was ~35 m. Some grew to a height that allowed the development of a slip face; the celerity (later) of medium-sized barchans was ~2 m day^{-1} (in high winds) (Elbelrhiti *et al.*, 2005).

The superimposition of dunes can also be the outcome of a collision between smaller, fast-moving, and larger, slower-moving dunes. A way in which this kind of superimposition could limit the size of dunes in a field of barchans is described in Chapter 4.

The Profile of a Fully Grown Dune

A dune is shaped by constant and rapid iteration between two processes: the entrainment and transport of sand (Chapter 1); and the momentum, velocity, pressure and other properties of the wind as it adjusts to the change in the shape of the dune after the sand has been removed. The nature of the adjustment of a moving fluid to changes in the shape of the surface over which it flows has been understood for over 150 years (since the full development of the Navier–Stokes equations). The breakthrough, as concerns dunes, came much later when these equations were built into a model of flow over smoothly rounded hills (Jackson and Hunt, 1975). The Jackson–Hunt model, in one form or another, is incorporated into most of the models discussed later in this chapter. Some of its elements have also been incorporated into computational fluid dynamic (CFD) models, which themselves have been used as confirmation of most of the mathematical models of the shape of dunes discussed in this chapter.

An important preliminary to any discussion of shape is to emphasise its link with the movement of the dune, the details of which are discussed later. Because movement is generated when some of the sand taken from the windward slope collects on the slip face, the form of each component, the trapping efficiency of the slip face (also later) and the movement of the whole are dynamically connected (detail also later). After the saltation cloud and ripple dynamics, this is the third example of a self-organising system in dune formation. Others follow, especially in Chapter 4.

The cross-section of a fully formed dune can be divided into eight parts (Figure 3.1).

Toe

The shape of the concavity at the toe is in constant adjustment, even to minor changes in the supply of sand and to the speed and direction of the wind, but one constant at the toe is a drop in velocity of the wind (some field data are shown in Figure 3.1). This is amply confirmed by the Jackson–Hunt model and CFD

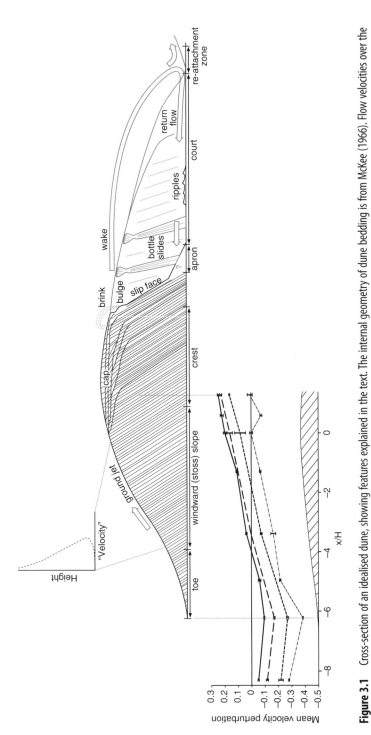

Figure 3.1 Cross-section of an idealised dune, showing features explained in the text. The internal geometry of dune bedding is from McKee (1966). Flow velocities over the windward slope are from Weaver and Wiggs (2011), thoroughly edited, collated and redrawn. The flow-velocity profile in the ground jet is highly generalised after Hunt *et al.* (1988, RL). The 'bulge' and the 'bottle slides' (Figure 3.2c) follow Anderson (1988). The bulge is very exaggerated. The form of the wake, reattachment point and return flow are taken from Parsons *et al.* (2004a). If there has not been a southerly wind just before the image was taken, then the dunes at 26°48′N; 30°01′E; 600 m, on Google Earth, show many of the morphological features in this figure.

models, and by observations in the field (some shown in Figure 3.1). The drop is a consequence of the backup of the oncoming wind against the dune.

Now, if the rate of sand transport is related only to wind velocity or shear velocity (u_*), as in Bagnold's model of sand transport (Chapter 1), the capacity of the wind to carry sand would be reduced at the toe where the wind decelerates. If, as a consequence, the wind were to deposit surplus load there, the dune would grow backward into the wind, which windblown dunes do not do. 'Anti-dunes', under fast-flowing water, do move upstream, but windblown anti-dunes have never been reported on Earth, although there may be windblown anti-dunes beneath the dense atmosphere of Venus (Chapter 9).

The explanation for this anomaly is almost certainly that sand transport at the toe is maintained by an intensification of turbulence, as, for example, measured by the Reynolds stresses (Chapter 1), and this too has been amply confirmed in the field and models of various kinds (Weaver and Wiggs, 2011; Figure 3.1). In the 'flow-curvature hypothesis', it is the concavity forced onto the path of the wind at the toe that intensifies turbulence (Castro and Wiggs, 1994). This hypothesis has some confirmation in the findings of a CFD model, which shows an increase in vertical velocity at the toe (Parsons *et al.*, 2004a), by another model of flow and erosion over dunes, and more field evidence (van Boxel *et al.*, 1999), but there are still doubts as to whether the curvature effect is sufficient to increase the level of stress enough to compensate for the drop in velocity (Wiggs, 2001).

Windward slope (or 'stoss slope')

The slope beyond the toe is slightly concave or straight to the crest. Its angle relates loosely to the height of the dune: on a dune that is 1 m high, it is between 1° and 4°; on a 5 m high dune, between 8° and 11°. The angle may reach 12° or 15° on higher dunes, but there is a great spread on a plot of height against average slope, probably because of differences in the grain size and wind regime (Momiji and Bishop, 2002).

In the Jackson–Hunt model, there are distinct flow regions in the wind above the windward slope, each with its own characteristics. The higher flow regions have little geomorphological significance, but close to the surface, the 'inner layer' or 'ground jet' directly controls the erosion and transport of sand. As shown diagrammatically in Figure 3.1, the ground jet is a shallow layer with a higher velocity than the layers above and below, which inverts the usual pattern in which velocity increases smoothly with height (Figure 1.1). Modelling shows that the velocity of the jet is very sensitive to the angle of the windward slope, the background value of u_*, changes in wind direction (even slight ones) and the character of separation in the lee (Parsons *et al.*, 2004a). If all else is constant, ground jets on steeper windward slopes are faster. The depth of the ground jet is directly related to the size of the dune. On a 6 m high dune, it is ~1 cm deep. It is so shallow, even on higher dunes, that saltating sand leaps through it. On small and medium-sized dunes, it is too shallow to be penetrated by conventional anemometers, although sonic anemometers may now be able to reach this layer (Weaver and Wiggs, 2011). Weaver and Wiggs's measurements of wind speed and turbulence over a barchan in Namibia showed that there was considerable acceleration of flow up the windward slope, even above the ground jet, and a corresponding decline in turbulence (Figure 3.1).

A simple geometrical argument shows that, to preserve the shape of the windward slope, the maximum rate of sand transport must be near the top, not at the crest (Bagnold, 1941, p. 183; Momiji and Warren, 2000). This has been confirmed both by field observations and by CFD models (Parsons *et al.*, 2004a), and is an important assumption in many mathematical models (shortly). In fact, it is inevitable, given the properties of flow over a mound of whatever material and in whatever fluid (Claudin and Andreotti, 2006). In wind speeds well above the threshold for sand movement, the peak speed has consistently been found to relate linearly to the height of the dune (reviewed by Momiji *et al.*, 2000). Peak velocity is reflected in peak sand flow: in a field study, the peak rate of sand transport in light winds had a ratio of 158:1 to the sand transport on the upwind plain, and a ratio of 13:1 in higher winds (Lancaster, 1985b).

The Jackson–Hunt model opened the way to mathematical modelling of the windward slope. It was now possible: (1) to model the distribution of shear stress over an approximation of the windward slope; (2) from this, to calculate the flux of sand, using the kind of models described in Chapter 1; and (3) to calculate the change in height at any point (by deposition or erosion). A first generation of models based on this approach aimed to produce an equilibrium or steady-state form, but none succeeded. In Howard and Walmsley's (1985) model, the angle of the windward slope continuously declines, and there is extreme sensitivity to small undulations. In Jensen and Zeman's (1985) model, in contrast, the windward slope continuously steepens, a tendency that cannot be countered either by introducing a correction of the transport rate according to slope angle, or by a lagged relation between wind speed and sand flux. Wippermann and Gross's (1986) model is based on a simpler flow model than Jackson and Hunt's, and produces a recognisable dune from a conical pile of sand, but the dune continues to shrink and lengthen. In a more recent model and using more complex models of flow, the height of the dune stabilises, but the windward slope continues to elongate (van Dijk *et al.*, 1999).

A second generation of models has been more successful. Unlike the earlier models that sought equilibrium, the first of these assumes that the system is already at equilibrium (in the sense that the amount of sand being added to the dune is the same as the amount that leaves it). In this model, as in nature, the windward slope steepens as the dune grows, and the angle of the windward slope is steeper at higher upwind u_*. The model prompts speculation that the steeper windward slope of high dunes is a morphodynamic adjustment: if the windward slope on a higher dune were steeper, shear stress would be greater, and more sand would be projected beyond its brink. In other words, the higher dune would have a similar sand-trapping efficiency to that of a lower dune, as will be seen shortly (Momiji and Bishop, 2002).

Sauermann and colleagues (2001) developed another second-generation model (labelled 'the Sauermann model' in what follows). An updated version of this model is described by Durán and colleagues (2010). It has three main inputs: a two-dimensional mound of sand in the shape of a Gaussian curve; the strength of the wind; and the input of sand (Chapter 1). The model is based on two crucial assumptions. The first is the peaking of wind speed before the crest, which has strong theoretical and empirical support (earlier). The second is less an observation of flow on the dune than of the behaviour of sand movement in the wind. It is the 'saturation length' or L_{sat} (explained in Chapter 1 and earlier in this chapter). The Sauermann model also assumes that the shape of the separation bubble in the lee (later) has an upper surface that is the mirror image of the dune downwind of the crest (a debatable approximation, Figure 3.2d). The model has been adapted by other workers (for example, Andreotti *et al.*, 2002a). Many of its predictions conform to field observations, and it has been used to produce (and test) hypotheses about the effects of different wind speeds and grain sizes (and other variables) on dune form (shortly).

A third second-generation model, built on some further realistic assumptions about the effect of the slopes (windward and leeward) on airflow and sand movement, has been developed by Pelletier (2009). The model combines Werner's (1995; Chapter 4) cellular automaton model, with a model of shear over dune-like shapes (as created by the cellular automaton), based on the Jackson–Hunt model. It achieves an equilibrium wavelength of ripples and dunes, unlike many others.

Crest

There are three main questions about the crest. The first and biggest is: why, on some dunes, is convexity so pronounced that the crest and brink are widely separated, both in downwind distance and in height? And while on others it is no more than a reduction in slope angle before the brink? And on yet others the windward slope is straight to the brink (crest and brink are coincident)? Convexity of the crest is more common in lower than higher dunes (Andreotti *et al.*, 2002a), but there are straight-sloped small dunes and convex crested high dunes. The second question is why, on many crests, is the convexity underlain by a deposit (or 'cap'), about 0.5 m deep, made up of sets of ripple strata, dipping gently downwind, separated by truncation surfaces (Figure 3.1)? The 'topsets' preserved in aeolian sandstones (Figure 10.3) are testimony to the universality of caps.

There are three sets of hypothesis for crest–brink separation. The Sauermann model assumes that the point of maximum shear precedes the point of maximum load (earlier), by about 10% of the total downwind length of the dune (Claudin and Andreotti, 2006). This allows it to produce a family of profiles of the crest, each with a different curvature and proportion of the windward slope in the upper convexity, and these are similar to profiles measured in the field (see also Parteli *et al.*, 2006). The cap may be explicable using this model (for example, if the growth of the cap is an adjustment to flow conditions), but its authors made no mention of it.

The second hypothesis invokes flow characteristics. Just as high turbulence boosts sand-carrying capacity at the concave toe, the drop in turbulence over the windward slope should reduce the sand-carrying capacity and favour deposition at the crest (van Boxel *et al.*, 1999). The decline in turbulence is now confirmed by field measurements by Weaver and Wiggs (2011), among others, although they believed the change in turbulence to be a minor element in the explanation for the shape of the crest. These models may explain the cap, but the process of cap formation has been neither modelled nor observed.

The third and simplest hypothesis invokes a changeable wind environment. Even the most directionally constant of winds go through a daily cycle of velocity and direction (linked to the Ekman model, Chapter 4) and it is the crest that responds first and most radically. Frequent and sometimes radical change in the crest has been shown by measurements in south-western Spain (Navarro *et al.*, 2011). The same kinds of observation have been made in dune networks (Chapter 4).

The models discussed here are not necessarily exclusive; they may act together, in diverse combinations, or each in different conditions.

Lee slope

Dome dunes

Dome dunes are defined here as those that have no slip face. This does not preclude movement: one set of dome dunes moved at $4\,m\,yr^{-1}$, in the same way as slip-face dunes, by erosion on the windward slope and accumulation in the lee (Bristow and Lancaster, 2004).

In an adapted version of the Sauermann model (earlier), dome dunes are found to be feasible, but unstable. They are liable to disappear if the sand supply drops below a certain threshold (because they lose sand more rapidly than do slip-face dunes); and if supply passes a higher threshold, they are transformed into slip-face dunes (Andreotti *et al.*, 2002b). The model shows an overlap in the size of dome dunes and slip-face dunes: the biggest feasible dome dune is ~0.8 m high, and the smallest slip-face dune is ~0.7 m high. One curiosity in the model is that the celerity (or rate of movement of a dune, later) of small dome dunes is independent of their size, but that celerity suddenly declines when the dune increases to a critical size, beyond which celerity is inversely proportional to the size of the dome, as in slip-face dunes. Most dome dunes in the model move more slowly than slip-face dunes of the same size. Dome dunes would almost certainly also occur at very high wind velocities (Momiji and Warren, 2000).

Figure 3.2   (a) Proto-dunes driven by a cross-wind on the lee slope of a large transverse dune; wind from the left. (b) The lee side of a small artificial dune (developed from a pile of sand), showing some of the features in Figure 3.1, particularly a brink, a slip-face and an apron. The windward slope is straight to the brink, whereas most dunes have curved crests; wind from the right. (c) A bottle slide on the slip-face of a dune. (d) Return flow in the lee of a dune visualised with smoke; wind from the right.

Slip-face dunes

Brink

Where the angle between the slope of the crest and the lee slope is steeper than ~14°, flow separates from the dune and is projected beyond it (Figures 3.1 and 3.2b). Most saltating sand does not follow the flow but travels a short distance beyond the break of slope before falling into the relative calm of the lee. The sharpness of the brink is due to the upslope growth of small avalanches, which end at the brink (shortly).

Bulge

On an actively accumulating slip face and in the right light, it is easy to see a bulge (Figure 3.1, where the bulge is greatly exaggerated), only a few millimetres to 1 cm proud of the straight, inclined surface of the slip face, and between 0.2 and 0.4 m downslope of the brink. The bulge is built by the greater intensity of sand-fall around that point than further up or down the slope. The concentration of fallout onto this position on the slip face is the result of two effects: first, the greater angle of the slip face than that of descending saltons; and second, the drag on particles falling into the gentle flow in the lee. Coarse grains fall nearer the brink; finer ones further out (Anderson, 1988). At common wind speeds, 84–99% of sand leaving the brink falls within 2 m of the crest, variation depending on differences in the length and shape of the windward slope of the dune. Field observations do not wholly confirm Anderson's model in respect of the position of the bulge, which has been recorded further downslope in the field than in the model (for many possible reasons) (Nickling *et al.*, 2002). A correction of Anderson's model of the formation of the bulge may remove some of the mismatch between field and theory (Momiji and Warren, 2000).

Slip face

Little of the activity on the slip face (as on Figures 3.1 and 3.2b) is driven by the wind, but it is described here because it completes the cycle of processes that move dunes (shortly). When sand falling on the bulge builds it up to a critical angle, the slope fails. Most failures occur when sand builds up to the 'angle of repose', but, if undisturbed, the angle may continue to steepen to a slightly higher 'angle of maximum stability' (or 'static angle of repose') at which failure always occurs. After a failure of slope at the bulge, a tiny scarp, 0.5–1 mm high, cuts rapidly upslope towards the brink. Grains tumble off the retreating scarp, feeding the avalanche. The scarp is active for many minutes before it halts on reaching the brink or when it merges into the slope. When and if the scarp reaches the brink, it may bring down somewhat coarser sand that has moved up the windward slope by reptation and creep (Anderson, 1988). The avalanche, fed by the tumbling sand, is constricted or 'bottlenecks' as it passes through the bulge, where the flow of sand is fastest (Figures 3.1 and 3.2c), hence the term 'bottle-slide' to describe the shape of the whole slide. After an avalanche, the bulge grows again to the critical angle, until it again fails.

The slip face is kept at the angle of repose. In dry sand, the angle of most of the slip face alters little with grain size. Most slip faces have slopes between 32° and 35°. On snow dunes, the angle is ~38°. It is remarkable that despite the variety in the size, sorting, shape and roughness of sands, the filling of the interstices between them, and of the way in which they are disturbed, that the angle of slip faces is so constant (Baas, 2007).

Below the bulge, an avalanche expands to a constant or slowly expanding width (Figure 3.2c). The retreat of the upper scarp feeds the avalanche for about a quarter of its duration, after which it becomes a series of waves moving downslope. Particle-image velocimetry reveals that collisions between particles are the main form of motion in the upper avalanche. Lower down, the movement is shock-like (frictional) or compressive (Tischer *et al.*, 2001). Tumbling brings coarser or platy particles (mica, shale, fragments of shell or diatomite, or organic debris) to the surface, most of which are moved to the sides of the avalanche, where, if enough, they form tiny ridges or 'levees'. Lateral expansion is caused by rebounds between moving grains (Goujon *et al.*, 2007). Avalanches are less than 1 cm deep and ~20 cm wide (Hunter, 1985a). The wind-transverse length of a slip face is scarred by several avalanches, each at a different point in the cycle (Figure 3.1).

Some slides never reach the base of the slip face, either because the supply from above runs out, or because a high proportion of the sliding sand is immobilised in the levees (Anderson and McDonald, 1990, RL). If an avalanche hits the base, a wave travels upslope at between 0.05 and $0.10 \, \mathrm{m s^{-1}}$ (Bouchaud *et al.*, 1994). The wave probably marks the limit between settled, stationary sand below, and looser sand above. All this activity creates a micro-relief on the slip face, which is constantly being smoothened by the passage of weak flurries of wind, which spiral upwards and laterally, driven by the upward momentum in this region. The frequency of avalanching on any one stretch of a slip face depends on the maximum u_* on the windward slope of the parent dune (Breton *et al.*, 2008).

The change in height of a slip face along its plan-shape curve (Chapter 4) has implications for trapping efficiency (earlier). On a dune in Nevada, ~15% of the sand discharged from the portions of the crest that were at an angle to the oncoming wind was blown along the curve by a lee-vortex (Walker, 1999). As the height of the slip face declines laterally, it reaches a point at which most saltating sand hops over it. The brink and the slip face then fade out laterally quite suddenly (Figures 3.1 and 3.2b; Bagnold, 1941, p. 211). The minimum size of a slip face on a particular dune can be predicted using Anderson's (1988) model. When ambient u_* is $0.5 \, \mathrm{m s^{-1}}$, and grain size is 0.25 mm (common values), projected particles reach no farther than ~1 m beyond the crest. This gives a minimum height, in these conditions, as 0.6 m. In high enough winds, eddies can move across the slip face and sometimes hold light debris for minutes at a time, but they are not able to have any material effect on the geometry of the lee slope itself (a hypothesis rejected long ago).

The slope of the slip face is steeper if the sand is damp or saline, because of greater cohesion between the particles (Chapter 1). A slip face that would have been at 35.6° when dry holds up at 43.36° with only 0.95% water in the sand (Carrigy, 1970). On the wetter margins of the desert, early morning dew may weld tablets of sand (up to a metre or more across), which may slide slowly downslope, breaking up as they do so (Schenk, 1983).

Trapping efficiency

The trapping efficiency is the ratio of the amount of sand falling on the lee slope to that bypassing it. One of the main controls on trapping efficiency is the maximum u_* on the windward slope. Other controls include the height of the slip face, the grain size and the ambient wind velocity. In gentle winds, the trapping efficiency grows rapidly with the height of the dune, but when the wind reaches a

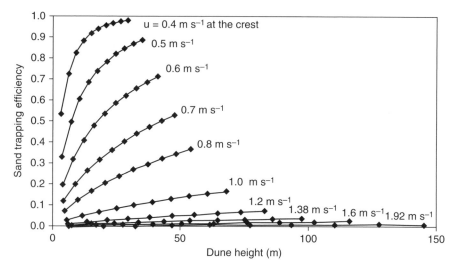

Figure 3.3 Trapping efficiency on dunes of different heights and at different wind speeds (Momiji and Warren, 2000). Reprinted with permission from John Wiley & Sons.

certain (high) velocity, all sand bypasses the lee slope, and trapping efficiency rapidly declines and virtually disappears (Figure 3.3).

Bedding

A succession of discrete slip face deposits or 'foreset beds' is created by avalanching sand as the dune moves forward (Figure 3.1 and shortly). These are known as 'translatent' strata or 'cross-bedding' if they are preserved between what were horizontal bounding surfaces at the time of their formation, or 'climbing translatent strata' if the bounding surfaces were not horizontal, these being a common feature of aeolian sandstones (Hunter, 1977b; Figures 3.1, 3.4, 10.3 and 10.4).

As with any sand that has been disturbed, the slip-face beds are 'reverse-graded': coarse grains are taken to the surface, leaving fine grains to make up the body of the avalanche. The outcome of the sorting depends on the size mix of sands delivered over the brink, the height of the slip face and the velocity of the wind at the brink (Kleinhans, 2004). The process creates a distinctively windblown pattern of 'pin-stripes', which, with the pin-stripe bedding created by ripples (Chapter 2), and also reverse-graded, are the best diagnostic of windblown dunes. Some ancient sequences show variations in the thickness and grain size of slip face beds, which have been interpreted as one would 'varves' in the bed of a lake: they record periods of different wind speed (Stokes, 1964). Particles in which there are heavy metals rise to the surface on a moving slip face. Where preserved in the bedding, a concentration of heavy metals is probably evidence of a windier period (Buynevich *et al.*, 2007).

Figure 3.4   Dune bedding revealed by wind erosion in weakly cemented calcareous aeolianite in the south-eastern Wahiba Sands in Oman (also discussed in Chapter 10). Image by Richard Turpin, RGS collection. Reprinted with permission from the Royal Geographical Society (with IBG).

Singing sands

Accounts of 'singing', 'sonorous' or 'booming' slip faces date from at least as long ago as the original stories in the Arabian Nights, and have attracted amateur and scientific attention ever since (including Bagnold, 1941, pp. 247–256). Quite natural, everyday slippage can set off the boom, but any small disturbance will do. Singing dunes are the exception rather than the rule, but they are common enough. Jebel Nakous in Sinai, parts of the Badain Jaran desert of China and the Kelso Dunes in southern California are among the most visited. The sound is generated only in very well-sorted, dry sand and is more or less independent of the mineralogy or shape of the grains. Measured frequencies are from 70 to 100 Hz, with some higher notes. The sound, which is independent of grain size, may 'last for several minutes, even after the avalanche has stopped' (Hunt and Vriend, 2010, p. 281; listen at http://www.sonicwonders.org/?p=1025). The effect probably comes from magnified vibrations in the very evenly sized cavities between well-graded sand grains, generated by the movement of an avalanche over immobile, more compressed sand beneath (Hunt and Vriend, 2010).

Apron

Some of the sand that is projected beyond the brink, as well as sand that is taken back towards the dune by counter-flow in the court (shortly), may build a low-angle wedge of sand against the base of the slip face, viz. the 'apron' (Figures 3.1 and 3.2b). Aprons are 1–2 m wide and ~0.5 m high. They probably play a very small role in dune dynamics, which may explain why little is known about them. In addition to sand projected over the brink, it is likely that changes in the direction of the ambient wind are responsible for aprons and that most are built of fine sand, being the only grain size that can reach them.

Court

The court (Figure 3.1) stretches from the foot of the slip face (or apron) to the reattachment zone. Beyond the brink, flow separates from the dune and is transformed into a 'wake', which gently rises to an altitude somewhat higher than the parent dune, where it picks up momentum from the ambient wind (Parsons *et al.*, 2004a; Figure 3.1), before descending to the reattachment zone (shortly).

At ground level in the court, flow returns from the reattachment zone towards the dune (Figure 3.2d). Its direction is reflected in ripple asymmetry and in the directions of accumulations in the lee of small obstacles like bushes. Wind speeds on the surface of the court have only ~17% of the free-stream velocity (Qian Guangqiang *et al.*, 2009b). CFD models, in general, confirm observations of courts in the field and in wind tunnels (Parsons *et al.*, 2004a). The character of the size and shape of recirculation bubbles is of interest in the construction of models of the shapes and spacing of successive transverse dunes (Chapter 5).

Three controls on the length of the court have been proposed:

- The height of the parent dune. In Schatz and Herrmann's CFD model, the length of the re-circulation bubble, downwind of a single dune, is four to six times its height (H) (2006). Another CFD model gives lengths of between $3H$ and $15H$ (Parsons *et al.*, 2004a). A field study of separation bubbles found a range between $4H$ and $10H$ (Walker and Nickling, 2002). Some modelling suggests that the length of the court is not directly proportional to H: big dunes have proportionately shorter re-circulation bubbles (Andreotti *et al.*, 2002a).
- Courts are longer in higher winds (Parsons *et al.*, 2004a).
- The cross-sectional shape of the parent dune. Courts are shortest on dunes with windward slopes straight to the brink, and on dunes with a pronounced downward, downwind slope before the brink; and longest on dunes with a slight reduction in slope before the brink but with no crest–brink separation (Parsons *et al.*, 2004a). A wind-tunnel study found that as the angle of the windward slope increased to 15°, as did the length of the court, and that circulation above the court was then best developed. Increases in the windward slope beyond 15° (which is unusual in reality) had no further effect. There is also a distinct jump in the energy of turbulence when the windward slope exceeds 15°, implying that some threshold has been passed; this may explain the scarcity of lee slopes above that angle (Dong Zhibao *et al.*, 2009c).

On the base of the court, as at the toe, turbulence plays an important part in entrainment and transport, so that even the weak reverse winds on the surface of the court can drive light debris and fine sand back towards the base of the slip face, or the apron (Walker and Nickling, 2003).

Reattachment zone

After reaching its maximum height, the wake curves back to the surface (Figure 3.1), following a path that approximates an ellipse, and may reach the reattachment point at an angle close to 90° (Schatz and Herrmann, 2006; Figure 3.2d). The momentum that has been acquired aloft is brought back to the surface in the reattachment zone, creating the fastest surface winds of the whole traverse. The flow splits in two: one branch back over the court; and another downwind. Turbulence is intense, and flow direction erratic, so that there is a high rate of sand transport, except in light winds. The zone is a few metres long (Walker and Nickling, 2003).

Where there is no second dune downwind of the first, shear grows quickly to a point about 12 dune heights downwind, and then more slowly to as far as 25 times the height of the dune. The boundary layer has then recovered the characteristics it had upwind of the dune (Walker and Nickling, 2003). This may explain the length of the sand-free courts of lone barchans, which reach 1 km or more downwind (21°33′N; 16°44′W; 3 km).

Movement

Movement is the consequence of erosion on the windward slope of a dune, and deposition in the lee (Figure 3.5). In high winds, dunes can move at a metre a day (for example, Hunter and Richmond, 1988). A dune, behind which Bagnold camped in 1931, had moved at 7.5 m yr^{-1} when located 57 years later (in total about 4 km) (Haynes, 1989). In Liwa in the United Arab Emirates, optically stimulated luminescence dating (Chapter 10) revealed that there had been 250 m of movement of a ~150 m high dune in 320 years (1.28 m yr^{-1}), with a burst of celerity 220–110 yr ago (Stokes and Bray, 2005). The highest celerities yet recorded are those of dunes in the eastern Bodélé depression in northern Chad (16°51′17″N; 18°22′28″; EA 1.7 km; located on Figure 10.11). The fastest dune there, which was no less than 25 m high, has been moving at ~20 m yr^{-1} (Vermeesch and Drake, 2008).

Bagnold (1941, pp. 203–205) used the word 'celerity' to denote the rate of movement of dunes. He proposed that celerity could be expressed as:

$$c = q / \gamma H,$$

where c is celerity in metres per hour; q is the sand discharge (in m^3 (m-width)$^{-1}$ s^{-1}), usually measured upwind; γ is the bulk density of sand (which Bagnold estimated to be 1.7); and H is the height of the dune.

The equation shows that higher dunes move more slowly, all else being equal. In general terms, this relationship has been amply confirmed by field measurement (Figure 3.6), with some exceptions, which will be discussed shortly.

A fuller equation for the movement of dunes, incorporating the height of the dune, the sand-trapping efficiency, grain size, air density, the gravitational constant and u_* (measured upwind of the dune), is given in Momiji and Warren (2000). In most cases (exceptions shortly), trapping efficiency is related directly to the height of the dune, and this is confirmed by Momiji's corrected version of Anderson's theoretical study of grain projection over the

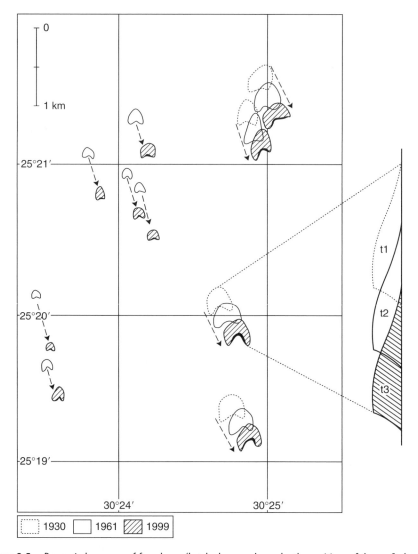

Figure 3.5 Downwind progress of free dunes (hatched areas, shown by the positions of dunes; Stokes *et al.* 1999); combined with a cross-sectional cartoon of the progress of three of the dunes. More recent positions of these dunes can be seen on Google Earth at about 24°49′15″N; 30°27′59″E; 10 km.

crest (earlier). The height of a dune and (upwind) u_* interact to produce a spread of dune celerities (Figure 3.7). The figure shows that there is virtually no movement in high winds for dunes of any height, because the trapping efficiency is then reduced near to zero. On Earth, gusts may reach beyond the velocities on the figure, but consistent velocities of that magnitude are

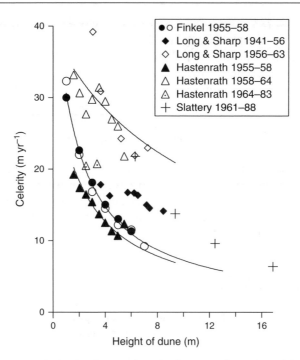

Figure 3.6   Celerity versus the height of moving dunes (Andreotti *et al.*, 2002b). With kind permission from The European Physical Journal.

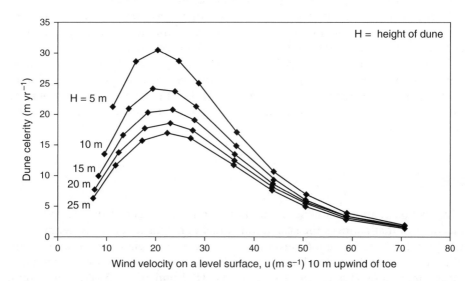

Figure 3.7   Dune celerity versus wind speed at 10 m above ground on an upwind level surface (Momiji and Warren, 2000). Reprinted with permission from John Wiley & Sons.

exceptional. A more recent and more general expression, also specifically for dunes, is given by Elbelrhiti and colleagues (2005).

Figures 3.3 and 3.7 answer some questions raised by field observations. First, why is there a rapid decline in the celerity of low dunes as wind speeds increase? Two answers have been proposed: (1) low dunes have a higher bulk density (Finkel, 1959); and (2) small dunes experience a more variable wind environment (Hastenrath, 1978; Dong ZhiBao *et al.*, 2000c). Second, why do the celerities of low and high dunes converge at high wind velocities, as noted by Sarnthein and Walger (1974), among others? Figures 3.3 and 3.7 show that both behaviours are functions of trapping efficiency: at high wind speeds, very low dunes have a very low trapping efficiency (they lose a high proportion of the sand that passes over them); on progressively higher dunes, the trapping efficiency becomes progressively less a function of height, as winds over the crest/brink are accelerated by the steepening of the windward slope (earlier).

Turnover time, bulk transport

Turnover time is the time between the entry of a grain into a dune and its exit from it. At its simplest, turnover time (T) can be defined thus:

$$T = L / c,$$

where L is the wind-parallel length of the dune, and c is its celerity.

Bulk transport is the transport of sand by dune movement (the rolling over of sand in the dune), as opposed to transport across the desert floor (Chapter 3) (Lettau and Lettau, 1969). One barchan in Morocco shifted no less than 1.2 million $t\,m^{-1}\,yr^{-1}$ or $20{,}550\,m^4\,yr^{-1}$ (Sauermann *et al.*, 2000). In a field of isolated dunes in Western Sahara, bulk transport was calculated to be $1.1\,m^3$ $(m\text{-width})^{-1}\,yr^{-1}$ (Sarnthein and Walger, 1974). In Mauritania, bulk transport by the dunes was of the same order as the interdune rate of sand transport (Ould Ahmedou *et al.*, 2007).

Size

This section examines a general question about the height of dunes. A distinct mechanism that controls the size of individual barchans in a field is examined in Chapter 4. There is a good relationship between the height, width and length of dunes, so that the term 'dune size' can be taken to refer to any one or to all of these dimensions (Lancaster, 1995a, figure 18.2 and pp. 475–476; and, more precisely, in relation to the Namib Sand Sea by Bullard *et al.*, 2011). The few exceptions do not invalidate the following arguments.

Two explanations for the size of dunes are easily dismissed. The first is that they are underlain by bedrock topography. This has been repeatedly been shown to be false (as most recently in relation to the mega-dunes in the Badain Jaran desert in central northern China; Dong ZhiBao *et al.*, 2004f). Besides, it is unlikely that the underlying topography would have such a regular pattern (except perhaps the pattern of wind-eroded longitudinal mega-yardangs; Goudie, 2007). Second, the size of dunes is related to the wind-energy environment. This, too, is unlikely, given that the biggest dunes in the world are said to be in the Badain Jaran in China, which is a low-energy wind environment (although it may not always have been so). It may be that repeated phases of fixation and reactivation contributed to the size of those dunes (Dong Zhibao *et al.*, 2004f). Another argument against the size/wind-speed explanation is a model that shows that dunes are reduced in high winds (Momiji and Warren, 2000).

Flow-hierarchy models

Wilson (1972c) proposed that flow structures in the planetary boundary layer (PBL) fell into three size categories, each corresponding to a different size of aeolian bedforms: small, responsible for ripples; intermediate, for dunes; and the biggest, for 'mega-dunes'. His proposal built on analogies with fluvial bedforms, which were thought at that time to be determined by turbulent structures in flowing water, where the strongest influence was the depth of flow (Allen, 1968). As they apply to ripples and small dunes, most of these models have been abandoned, as explained in Chapter 2 on ripples and in Chapter 4 on transverse dune patterns.

Wilson's proposal that 'mega-dunes' were linked to atmospheric patterns of flow, however, has been revived. It is now claimed that the thermal inversion above the PBL acts like the surface in a stream, such that developing windblown dunes 'excite surface waves' on the inversion, which in turn modify the dunes, until they reach a size that is limited by the 'resonant condition' (Andreotti *et al.*, 2009, p. 1120). The spacing of large dunes would thus be determined by the depth of the boundary layer, itself dependent largely on temperature, such that it may be 3.5 km in the desert, but only 300 m on a cool coast. In deserts, the interaction could take place only in summer, because the cool atmosphere of winter is stable down to the ground. Thus (on Earth) mega-dunes are probably the subaerial dynamic equivalent of subaqueous (river, estuarine and submarine) dunes, while subaerial dunes are probably the dynamic equivalent of subaqueous ripples (earlier). A similar, though less well supported, model of transverse dune formation has been suggested for ice dunes in Antarctica (Chapter 4).

Grain-size models

In addition to his flow-hierarchy model, Wilson (1972a) proposed that the wavelength of dunes (ripples and mega-dunes) is linked to the grain size of their sand. Despite the flimsiness of Wilson's evidence, some recent measurements confirm the relationship as it applies to small 'proto-dunes' (Andreotti *et al.*, 2010). These observations have been supported by a model developed by Pelletier (2009), who left the door open to further examination of the effects of grain size on the shapes and sizes of transverse dunes.

The time/supply model

In this model, the size of a dune is a function of time and the regional sand transport rate (itself related to rate of supply and the ambient wind speed) (Warren and Allison, 1998). There is some support for the model in the heights of large and small dunes in the Gran Desierto in north-western Mexico and in the Namib Sand Sea, which appear to be proportional to local sand supply over different time periods. Size may also be the product of particular directional wind regimes: the dunes in the northern Namib Desert may owe their great size to a wind regime that takes sand from diverse sources and concentrates it there (Lancaster, 1988b; Livingstone, 2003).

These models may each apply at some times and in some places, or collaborate to increase the size of dunes. In the Namib, there may be collaboration between the time/supply hypothesis and the concentration of sand in a centripetal wind-directional regime. The time-supply model and Andreotti's model are also not at odds.

References

Anderson, R.S. and McDonald, R.R. (1990) 'Bifurcations and terminations in eolian ripples, AGU 1990 fall meeting', *Eos, Transactions, American Geophysical Union* 71 (43): 1344.

Hunt, J.C.R., Leibovich, S. and Richards, K.J. (1988) 'Turbulent shear flow over hills', *Quarterly Journal, Royal Meteorological Society* 114 (484): 1435–1470.

Part Two
1000 to 10,000 m^2; 100 to 1000 years

Chapter Four
Pattern in Free Dunes

Definitions

Part II of this book (in which this is the first chapter) adds the third spatial dimension to the description and understanding of dunes. Thus, 'pattern' in this chapter means three-dimensional shape, in contrast to the two-dimensional form of a single dune of Chapter 3. 'Pattern' includes both the three-dimensional shape of individual dunes and the pattern of groups of dune.

The chapter is about prototype patterns, defined as those: (1) whose essential features are replicated over extensive areas; (2) that persist over long periods of time, despite a continuous throughput of sand (even, in some cases, despite a change of winds); (3) that are composed of elements that maintain their essential characteristics and relationships as they move downwind (in transverse dunes), are elongated (in linear dunes) or grow upward (in star dunes); (4) that occur in some extraterrestrial atmospheres, as on Mars, Venus or Titan (Chapter 12); and (5) that are active, and without a history beyond the time they take to form. The later history of individual dunes and of assemblages of dunes is for Chapter 10.

The central assumption in this chapter is that wind directionality is the primary control of the differentiation of pattern, an assumption that is shared by most of the literature on dune pattern. The other boundary conditions that influence dune pattern (volume, disposition and temporal pattern of the delivery of sand; antecedent conditions; and the shape and the size of the accommodation space; Ewing and Kocurek, 2010a) are assumed here to be tertiary and more applicable to real dunes than to prototypes. As such, they are topics for Chapter 10.

Dunes: Dynamics, Morphology, History, First Edition. Andrew Warren.
© 2013 John Wiley & Sons, Ltd. Published 2013 by John Wiley & Sons, Ltd.

Wind-Directional Regimes

Global winds

On Earth, three wind systems have global or continental reach (Figure 4.1): the Hadley circulation, driving the Trade Winds in the mid- to low latitudes; the Westerlies in the temperate latitudes; and the Monsoons in India, China, western and southern Africa, northern Australia and southern North America (also shown on Figure 4.1). The Westerlies and the Trades combine to create wind gyres in central Australia and the Sahara, in both of which there are signs of half of a second gyre to the west of the main gyre (Figures 8.1 and 8.5); southern Africa contains only the western arc of a gyre. Like the major oceanic gyres, the winds flow clockwise round the continental gyres in the northern hemisphere and anticlockwise in the southern hemisphere.

Local wind systems

Sea breezes

These are driven by thermal contrasts between land and sea. They blow landward in the afternoon and seaward at night. They are strongest in the warm season, as on 80% of summer days on the Israeli coast (Halevy and Steinberger, 1974, RL). Figure 4.2 shows that the daytime, landward breeze is much stronger than the night-time seaward breeze; and that at any place, the landward wind starts quite suddenly, often behind a front, as in the UAE (Remiszewska *et al.*, 2007). Sea-breeze systems may reach 300 km from the shore, but winds with sand-moving capacity seldom reach as far.

Mountain winds

'Katabatic' winds occur where cold and therefore dense air in high country drains down valleys. On Earth, the most intense katabatic winds flow off the Antarctic Plateau, where they reach velocities that create gravel dunes (later). Katabatic winds also blow off the north polar ice cap on Mars (Chapter 12). 'Anabatic' (upslope) winds are driven by the heating of air in valleys and are usually weaker, although some can generate small dunes.

Some katabatic winds, take part in a diurnal reversing-wind regime with anabatic winds. The directional pattern of such a reversing wind system is shown in Figure 4.3f. In other places, reversing wind systems couple katabatic or anabatic winds with global wind systems. This is the case in the Great Sand Dunes National Park in Colorado. Other mountain winds, termed the Föhn in Europe, Chinook in North America, and Berg Wind in southern Africa, work at a larger scale and create strong winds on the lee sides of mountain ranges. These winds operate only in certain synoptic situations. Mountain winds create their own reversing patterns; other local winds combine with global winds to create other reversing regimes.

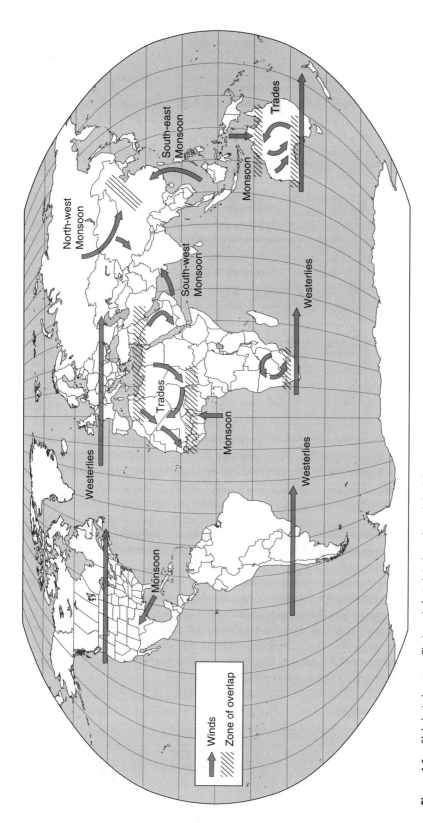

Figure 4.1 Global wind systems. The 'gyres' of winds in Australia and the Sahara are shown by dune patterns, as in Figures 8.1 and 8.5. The 'zones of overlap' are areas in which there is both seasonal alternation of the directions major wind systems (discussed shortly); and secular variations in wind pattern on scales of tens of thousands of years (or more), as explained in Chapter 10.

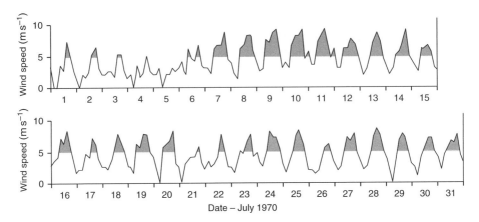

Figure 4.2   Daily variations of wind speed in a sea-breeze system on the Gulf coast of Texas at Corpus Christi. The stippled areas represent winds over 5 m s^{-1}, the approximate threshold of sand movement (Hunter and Richmond, 1988).

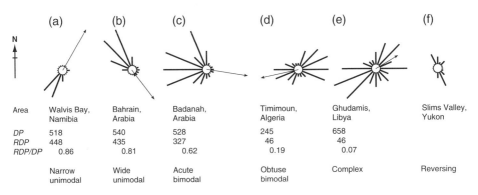

Area	Walvis Bay, Namibia	Bahrain, Arabia	Badanah, Arabia	Timimoun, Algeria	Ghudamis, Libya	Slims Valley, Yukon
DP	518	540	528	245	658	
RDP	448	435	327	46	46	
RDP/DP	0.86	0.81	0.62	0.19	0.07	
	Narrow unimodal	Wide unimodal	Acute bimodal	Obtuse bimodal	Complex	Reversing

Figure 4.3   Classification of directional wind regimes, from Fryberger and Dean (1979), with an added category (reversing). The arrow indicates the RDD (explained below). The data in the figure have been computed in the following sequence: (1) data on wind strength and direction from a meteorological station are converted to sand transport rates, using a modified form of the relationship of Lettau and Lettau (1978) (Chapter 1); (2) these data are used to compute (i) the total amount of sand that is blown in an average year from all directions, at each station, termed the 'drift potential' (DP); and (ii) the volume of sand blown from different directions (usually eight), termed 'vector units' (VU); (3) the VUs are used to calculate 'the resultant drift potential' (RDP) and its direction, the RDD; (4) the RDP is divided by the DP to give the index of directionality, with values ranging from 0 to 1, where '1' is a perfectly unimodal annual regime (winds from the same direction throughout an annual cycle), and '0' indicates a complex annual regime (winds from all over the place).

The Ekman Dune Effect

This is not so much a wind system, as a control on the direction of local winds. Ekman's model predicts that a wind blowing over a change of surface from smooth to rough swivels up to 35° from its angle of incidence at the change, to the left in the northern hemisphere and to the right in the southern hemisphere (explained in Markowski and Richardson, 2010, pp. 81–82, RL). Deflection is greater on rougher surfaces and at higher wind speeds.

The impact of this deflection is most marked on coastal dunes, where onshore winds pass from smooth sea to rough land (Warren, 1976a). In the northern hemisphere, the effect occurs in the Guerrero Negro dunes on the west coast of Baja California (28°03′70″N; 114°08′51″W; 850 m), where the direction of the incoming sea breeze is shown on the image by sand trails behind low obstacles at the head of the beach; when the wind meets the rough surface of transverse dunes, it swings to the left (at right angles to the slip faces). In the southern hemisphere, on the Peruvian coast (14°52′S; 75°31′W; 2 km), lee dunes behind hillocks near the coast show a south-easterly wind, which then swings sharply to the right when the going gets rough over transverse dunes further inland. On King Island between the Australian mainland and Tasmania (map in Jennings, 1957b; located on Figure 8.1), parabolic dunes (Chapter 6) both on the east and on the west coast, swing to the right as they move inland.

The 'Ekman dune effect' is not as easy to detect inland, where contrasts in roughness are less distinct. However, the effect is seen in three examples. The first, in the northern hemisphere, is a comparison of results from two towers of ane-mometers, one on smooth alluvium and another on a very rough lava field in the Mojave Desert. This being the northern hemisphere, the dominant wind dutifully swung to the left over the rough lava flow (Greeley and Iversen, 1987). The second, in the southern hemisphere, is on the Skeleton Coast in Namibia (19°48′S; 12°59′E; 16 km), where the onshore south-westerly first meets a smooth land surface, over which it is barely deflected, as shown by sand streamers, as at Guerrero Negro. Also, as at Guerrero Negro, the Skeleton Coast wind then blows over the much rougher surface of transverse dunes, where it swings right (passing over them at right angles); yet further inland, when the wind again meets smooth country, it swings back to the left, as ever shown by sand streamers (Lancaster, 1982a). The third inland example of the Ekman dune effect occurs in palaeodune patterns in the Sudan (13°53′N; 28°34′E; 8 km), where the north-nouth-easterly palaeowind was first swivelled to the left over the rough surface of large transverse dunes and then swung back to the right over the smoother surface of the lower transverse dunes (although there are other possible explanations for these particular patterns).

The Classification of Wind-Directional Regimes

Seasonal rhythms in wind direction brought about by the interaction of global, continental and local wind systems create many different wind-directional regimes. The only classification of these regimes for use in studying dune patterns is shown in Figure 4.3. The Fryberger-and-Dean method has its problems. First, it uses a single linear regression of the rate of sand drift against the drift poten-tial, whereas in data from Kalahari, and doubtless elsewhere, the slope of the

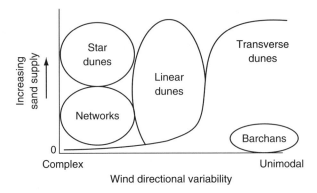

Figure 4.4   Plot of the directionality of wind regimes against a generalised measure of the availability of sand for dune-building (Livingstone and Warren, 1996). Dunes of different pattern fall into distinct zones on the plot (as explained in the text).

regression differs from station to station (Bullard, 1997a). Second are its frequency and magnitude biases. Third is its choice of wind-speed frequency categories (Pearce and Walker, 2005). None of these affects the essential conclusions of this chapter.

Wind-Directional Regimes and Dune Pattern

Wasson and Hyde (1983b), following a suggestion by Hack (1941), plotted the Fryberger-and-Dean index of directionality against the availability of sand (they estimated this from the smoothed depth of sand at a site, derived from dune heights and spacings), on a plot with the same axes as in Figure 4.4. Modifications to the Wasson-and-Hyde diagram were made first by Lancaster (1995a), and then by Livingstone and Warren (1996; Figure 4.4).

The key role of wind regime in pattern formation, shown on Figure 4.4, has now been strongly supported by two types of model. The first, flume experiments using a turntable, have shown that bedforms (windblown ripples and dunes, subaqueous dunes, etc.) obey the following rule: in directionally changeable flow regimes, bedforms orient their crests to maximise sediment transport (Rubin and Ikeda, 1990). The second type of model is the cellular automaton (Figure 4.5). Both are discussed shortly in reference to different dune patterns.

Transverse Dunes

Some transverse dunes are lengthy continuous ridges, as in Guerrero Negro (coordinates earlier); others are discontinuous, which, in the extreme, produces barchans (later). Some transverse dunes are little more than 0.5 m high; others

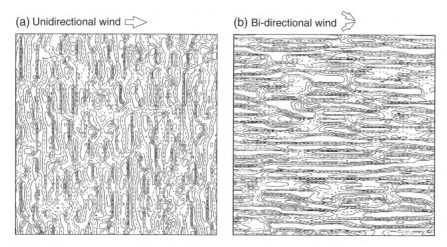

Figure 4.5   Transverse and linear dunes as modelled by a cellular automaton (Bishop *et al.*, 2002).

are over 100 m high, as at 26°45′N; 13°32′E; 12 km in central Libya, and in the palaeodunes of the Nebraska Sand Hills (42°07′N; 102°06′W; 33 km).

Transverse dunes occupy the whole of the lower portion of Figure 4.4 (where dunes are small), because small free dunes, having little mass with which to buffer change, are constantly being reorientated by winds from different directions, but retain their transverse form. In cellular automata (later), transverse dunes persist in regimes with 'winds' from two directions, if they are no more than ~45°, apart, at which angle there is a smooth transition to another pattern (for example, Werner and Kocurek, 1997). In the flume study of Reffet and colleagues, the pattern flips suddenly from transverse to linear when winds diverge by more than 90° (2010). The same critical angle is shown by a numerical model that incorporates a cellular automaton (Parteli *et al.*, 2009). This behaviour strongly suggests a degree of morphodynamics (feedback between form and process).

Two-dimensional pattern: vertical and downwind

The transverse pattern can be reproduced in very simple mathematical models. One such has three starting configurations: dunes all of the same height and spacing; dunes with the same spacing but different heights; and dunes with differing heights and spacings (Parteli and Herrmann, 2003). Movement of these 'dunes' and the sand flux over them is proportional to their height (more over bigger dunes). Sand moves from dune to dune so that their heights and spacings can grow or decay. The heights and spacing of the dunes converge over time, whatever the starting configuration. When the possibility of the coalescence of dunes is introduced, the heights still converge, and the total number of dunes declines quickly. When the parameter that incorporates wind speed (and grain size) is varied, it is seen to control the speed of evolution of the system. The model compares fairly well with some real transverse dunes at Lençóis Maranhenses in Brazil (2°33′S; 42°30′W; 15 km).

Early speculation about a one-to-one linkage between atmospheric turbulence and the downwind spacing of transverse windblown dunes and subaqueous dunes has been replaced by models of a two-way relationship, between fluid turbulence and the growth of the dune. One of these models suggests that at high wind velocities, the spacing of dunes is proportional to the saturation length (Chapter 1), itself dependent on grain size among other things. Another model shows that transverse dunes, in particular barchans (later), could behave like solitary waves or solitons: when a small dune catches up with a large dune (by processes explained earlier), the smaller moves through the larger, emerging on the far side, unscathed (Schwämmle and Herrmann, 2003). This claim has been firmly rebutted by field geomorphologists (Livingstone et al., 2005) and questioned or rejected by other modellers (Herrmann and Sauermann, 2000; Elbelrhiti et al., 2005), although the phenomenon appears to have been observed in a 45-year record on satellite imagery of the Bodélé Depression in northern Tchad (Vermeesch, 2011).

None of these models covers an uncommon, yet not rare anomaly: transverse dunes that have interdunes that extend well beyond the reattachment zone. One significant feature of these long interdunes appears to be a restricted sand supply, which may allow wind speeds to increase beyond the reattachment zone (Baddock et al., 2007).

Two-dimensional pattern: horizontal and transverse to the wind

In this dimensional mix, the pattern is of concave sections opening downwind (termed 'barchanoid'), which are higher than, and alternate with, lower convex sections (termed 'linguoid') in which the curves open upwind (Figure 4.6; the pattern is shown very well on those dunes at Guerrero Negro, 28°07′48N; 114°06′18W; 850 m). In general, cross-wind curvature is proportionate to height, but some small transverse dunes curve gently in this dimension; some have pronounced curves. At Guerrero Negro, and in many other fields of transverse dunes, the barchanoid sections are much longer across the wind than the linguoid. On most barchanoid sections, crest and brink are separate, whereas they are coincident on most linguoid sections (in which windward slopes are virtually straight from toe to brink). There is little evidence of how this system of dune shapes migrates, except at White Sands in New Mexico, where it appears, surprisingly, that sinuosity migrates at right angles to the general direction of dune movement (Kocurek et al., 2007). The broad outlines of the pattern are replicated in all cellular automata, but all fail to reproduce the differences between the cross-sectional shapes of barchanoid and linguoid sections.

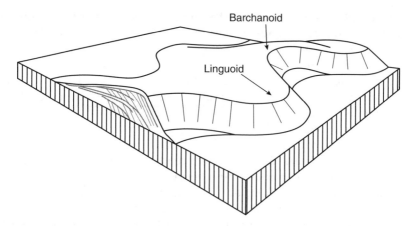

Figure 4.6   Cartoon of the three-dimensional pattern of transverse dunes, and associated terminology.

Self-organisation

The understanding of patterns in three dimensions has changed radically in the last decade and a half (the greatest change is in regard to transverse dunes, but the general perception of the development of other patterns has also changed). The main stimulus has come from cellular automata.

Cellular automata simulate the movements of slabs of sand (smaller than the dunes, but much larger than a grain of sand) in computer space, according to a set of rules. The choice of slabs, rather than individual grains of sand, is a restriction imposed by computing time: to follow individual grains in computer model would be impossible; quite small cellular automata simulating the movement of slabs already need days of computing time. The models begin with a gridded surface in computer space, on which there are small random differences in height. The surface may be underlain by deep sand (many slabs), simulating a large supply of sand for dune building; or by only a few slabs over an immobile substrate, simulating a system with less available sand. Slabs may or may not be fed in to the upwind grid and extracted downwind. Simulation begins with the random choice of a slab, which is moved downwind according to probabilities (as of distance moved and direction). If the slab lands on a patch that is above the general level, it remains there, following Bagnold's model of the growth of a patch to a dune (Chapter 3). In most models, a slab that lands on a down-current slope that is steeper than the angle of repose of dry sand 'slides' downwards. Successive slabs, chosen at random, are then moved and come to rest by the same rules.

Several improvements have been made to the pioneer model of Nishimori and Ouchi (1993a). Werner (1995) introduced separate timescales for downwind and downslope (slip face) movements. Momiji and colleagues (2000) introduced a simulation of speedup at higher altitudes on the dunes. Pelletier (2009) linked his cellular automaton to a modified version of the Jackson–Hunt flow model (Chapter 3), ensuring as well that the downwind length of a dune (assumed constant in the Jackson–Hunt model) could increase over time, which meant that the depth of the ground jet could increase also (Chapter 3); Pelletier made some further modifications. Some of his findings challenge the assumptions of earlier models: he found that the roughness length (z_0) on the desert floor had a strong effect on bed shear over the dune (suggesting support for Wilson's model of the relationship between the spacing of dunes and grain size; Chapter 3). In contrast to earlier cellular automata, the Sauermann model (Chapter 3) and most others, Pelletier's dunes reached an equilibrium height.

The first revelation in cellular automata is usually their reproduction of so many features and behaviour of real dunes, such as the relationship between celerities and dune size. Other similarities can be seen in Figure 4.4. Cellular automata are also strong confirmation of models of dune pattern.

The much more fundamental revelation of the automata is that dune patterns, like many other complex environmental (and social) phenomena (including systems of saltating grains, Chapter 1, and windblown ripples, Chapter 2), are self-organising systems in which a pattern emerges through the interaction of several processes. Although the understanding of both the individual processes and their interaction 'is in its infancy' (Kocurek et al., 2010, p. 52), some of the contributing processes are easily identifiable, being well known from field data. They include: (1) the relationship between rate of movement of dunes and on their size (Chapter 3); (2) the rate of supply of sand and its spatial variation, best seen in the comparison of barchans, which form where sand is scarce and transverse dunes when sand is abundant, but also in the streams of sand that leave the loose ends of transverse dune (termed 'defects', shortly); (3) the downwind repulsion of dunes by the circulation bubble (Chapter 3); and, somewhat at odds with the last item, (4) the regulation of the size of the largest barchan by off-centre collision by smaller barchans (shortly in relation to size regulation in barchans).

Another revelation of cellular automata is that pattern develops gradually. Until the ready availability of remotely sensed images, and of long enough time series of these images, studies of pattern development in real dunes had been possible only of the earliest stages, as on Padre Island (Chapter 3). Now that longer-term change is easier to study, there will be a surge of activity in this area. Some studies have already been published. One, of White Sands New Mexico, analysed three parameters of pattern: crest-length, spacing, crest orientation, and defects (terms defined on Figure 4.7; Ewing and Kocurek, 2010b). The parameters are related in the following way: first, more defects mean shorter crests; second, if defects from one dune ridge connect up to another, the result is an increase in spacing; and third, the overall orientation of a dune system is altered if defects differ in direction from the main dunes (as most do).

The analysis of patterns in the White Sands dune field found a range of behaviours including: merging (as one dune catches up on another); lateral

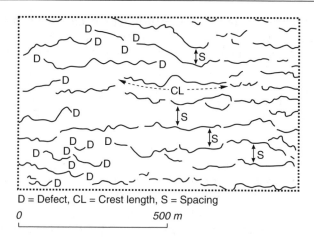

D = Defect, CL = Crest length, S = Spacing

0   500 m

Figure 4.7 Some measures of pattern, illustrated by transverse dunes at 23°03′52″S; 14°32′09″E; 26 km, near Walvis Bay in Namibia.

(cross-wind) linkage; defect repulsion (the loose ends of dunes failing to merge with dunes downwind); off-centre collision, as when the 'horn' or 'defect' of a dune merges with the lateral edge of a dune downwind; defect creation, when a dune breaks apart laterally; and some others. These interactions can create a trend either towards greater organisation, for example towards fewer, and therefore more widely spaced dunes or, strangely, towards poorer organisation (Ewing and Kocurek, 2010b). Pattern analysis is re-examined as a means of dating dunes in Chapter 10.

Barchans

Barchans are discontinuous transverse dunes, in which, because of a shortage of sand supply, only the barchanoid (defined earlier) element remains. In the Turkic (or Ural-Altaic) family of languages barchan means, simply, 'a free dune' (Burt, 1949, RL), and this meaning is retained in the Russian literature. It was also the usage in some late 19th- and early 20th-century accounts in other European languages, but Pompeckj (1906) in German, and Beadnell (1910b) in English, attached the name to an isolated, crescentic, free dune. This meaning, which was adopted by Bagnold (1941, p. 223), has persisted in the Western literature, and is the choice here. Views of barchans, thus defined, but with different shapes, sizes and patterns, are at 21°26′N; 16°52′W; 5 km, in Western Sahara, and at 16°47′S; 71°49′W; 2 km in southern Peru. The most frequently used term for a downwind extension of the flank of a barchan is 'horn' (Bourke and Goudie, 2009). Several possible explanations for these differences are discussed later, although there is still no definitive answer.

Barchans with heights greater than ~100 m have been termed 'mega-barchans' (as at Pur Pur in Peru: 08°24′04″S; 78°51′19″W; 1 km), although there is no precise size for the differentiation. In a survey of dune topography in eight locations using data from the Shuttle Radar Topography Mission, the mean mega-barchan spacing was 2.08 km, and the mean height was 68 m (Blumberg, 2006). Most large barchans are covered and surrounded by smaller transverse dunes (as at Pur Pur, coordinates earlier). The term 'mega-barchanoids' has been applied to groupings of smaller barchans or transverse dunes in a barchan-shaped pattern, whose height is no greater than that of the individual dunes (Kar, 1990; as in north-Chad: 16°51′N; 19°40′E; 12 km). Their genesis is mysterious.

Symmetrical barchans occur only where the annual change in wind direction is less than 15° (Tsoar, 1984). With this or a similar assumption, cellular automata create symmetrical barchans. But there is a wide range in the shapes of barchans, on both Earth and Mars (Bourke and Goudie, 2009). Barchans that have one longer horn (as at 24°06′N; 01°21E; 2.5 km in south-western Algeria) have been explained as the result of an asymmetry in the directional pattern of the wind (perhaps only temporarily), off-centre collisions between barchans and inclined underlying topography (Bourke, 2010).

Despite being one of the rarest of the dune prototypes, barchans have attracted almost as much research as all the other prototypes together (certainly in the last decade or so). There is more to this imbalance than the attractive shape and accessibility of these dunes. E. Guyon is quoted as saying 'barchans are our drosophila', in short, start small and simple (Andreotti et al., 2002b).

One of the early foci of research was the geometry of groups of barchans, which were found to have remarkably consistent geometry within an individual field, but different for different fields, strongly implying a shared response to prevailing conditions of wind, sand grain size and sand supply (Sauermann et al., 2000; Andreotti et al., 2002b). There are three questions about this uniformity.

The first concerns the geometry of single barchans. The preservation of the three-dimensional shape as the dunes migrate (and by extension the migration of the same shape component in transverse dunes, earlier) was first examined by Howard and colleagues (1978), who noted that sand avalanching down the curved, funnel-like plan shape of barchan slip faces would, unless compensated for, serve to increase the height of the central dune, distorting rather than preserving the shape of the dune.

Howard and colleagues also speculated that the lateral transport of sand on a barchan might compensate for the concentration, noting also that the cross-slope orientation of ripples on the flanks of barchans suggested the centripetal movement of sediment. This mechanism has now been modelled by Hersen (2004), whose model included the size-fractionation of grain sizes on the three-dimensional curves of the windward slope, and the outward movement of coarse, reptating sand. Hersen's model differentiates between the direction of saltating grains, which deviates only slightly from direction of the ambient wind, which itself is only slightly deflected as it passes over the curved windward slopes of the barchan; and the direction taken by the coarser reptating sand (Chapter 1), which, because it moves close to the surface, is very much less affected by the direction of the wind and is therefore more responsive to gravity. The deflection of coarse reptons to the horns reduces the celerity of the dune as a whole. Hersen's model shows that a barchan reaches an equilibrium shape only when the lateral movement of reptons is allowed for. If they were of the same size of sand as the main dune, the horns, behaving as small dunes, would move faster than the higher parent dune and destabilize its overall shape.

Hersen also calculated the effect on the shape of different proportions of coarse grains. With little coarse sand, a barchan has steeper slopes, a tighter curve and long horns (Figure 4.8). With more coarse sand, a barchan has gentler slopes, a wider curve, and shorter horns. Hersen found other controls such as on curvature and slope (not discussed here). As regards sand grain size and the shape of barchans, Hersen's model may explain some (but only some – see later) of the range in the shapes barchans: horns are short on some of the Martian barchans; slightly longer in the Peruvian barchans (coordinates given earlier); and wider and longer in the western Sahara (Bourke and Goudie, 2009). Bourke and Goudie speculated that the availability of sand might be another control on the shape of barchans. The size of the sands in barchan horns is not well documented,

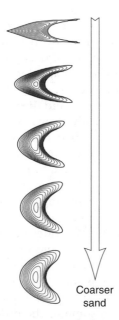

Figure 4.8   Effect on the shape of barchans of varying a parameter in a model that relates to the grain size of the sand (Hersen, 2004). With kind permission from The European Physical Journal.

but the few data there are show them to be coarse (Hastenrath, 1967). Hersen's model may also explain the shape of zibars (later).

The second question about the uniformity of shape (and size) of barchans in a group is: why, if barchans collide, as they do (shortly), do they not eventually amalgamate into one huge dune? The puzzle is compounded by the observation that the width of the horns is relatively smaller on large than small barchans. Because sand is lost from a barchan only via the horns, the large barchans lose proportionately less sand (Hersen *et al.*, 2004).

The first control on unchecked growth may be that large barchans loose proportionately more sand in storms (Elbelrhiti *et al.*, 2005). The second control on infinite growth may lie in their behaviour as a group: when a small, fast-moving barchan collides 'off-centre' with a slower-moving bigger one (a form of collision that is more common than full-on collision), there is erosion of one side of the windward slope of the bigger dune in the turbulent wake of the incoming dune (Chapter 3). The sand released in this way travels towards the horn of the bigger barchan and is lost to it either in enhanced discharge from the horn or in one or more small, fast-travelling new barchans. Hence, off-centre collision is a mechanism (not necessarily the only mechanism) that keeps the size of the biggest barchans from growing indefinitely (Hersen and Douady, 2005; where the supplementary material has videos of this behaviour in a flume). The implications of this behaviour have implications for pattern development (later and Chapter 8). Several variations on the theme of the collision of small with big barchans can be seen at 27°52′56″N;12°43′10″W, 4 km.

The third question about the shape (and size) of barchans is perhaps the most obvious: why are they so different in different fields? There are sharp distinctions that go beyond the explanations given so far, even in contiguous dune fields: compare the sizes and shapes of barchans in two neighbouring dune fields on the same Google Earth scene at 27°40′28″N; 13°08′53″W, 2 km in southern Morocco/Western Sahara, and the very different shapes of barchans in southern Peru (coordinates already given). An investigation of the differences between these three dune fields ruled out topography, granulometry, wind direction and speed (although the Peruvian wind data were poorly related to the dune field), and sand flux as explanations. The investigation was left with three possible explanations. The first was the observation that contiguous fields of barchans of different morphology originate from indentations in an upwind scarp, themselves of different size. The second explanation might be that there were differences in the substrates of the different fields of barchans (as between rough stony ground and a smooth salt pan). However, differences in shape or size are usually eliminated by merging and such like processes further downwind. The third explanation could be the occurrence of occasional strong winds coming from directions different to the prevailing wind, an explanation that might apply to some of the different shapes in the Peruvian fields of barchans seen in air photographic or satellite images taken at different times. Elbelrhiti and colleagues left open the possibility that differences in shape between dune fields are connected with the very much lower concentration of dunes in the Peruvian field than in either of their Moroccan examples (Elbelrhiti *et al.*, 2008).

A mathematical model of the pattern within a field of barchans provides some more insights into the interaction between dunes. It starts with a random distribution of dunes and then iterates the following processes: (1) new dunes, of specified sizes, are added upwind; (2) the position of all the existing dunes is recalculated according to their rates of movement and of their capture or loss of sand, allowing also for coalescence (earlier); (3) the interdune flux is recalculated according to the new dune positions, giving new rates of capture or leakage of sand to and from each dune. The model reveals: a decreasing density of dunes downwind (as in the Laâyounne, the prototype); fluctuations in the number and size of the dunes, but about a steady mean; and a remarkable constancy in the width of the field, again as in the prototype (Lima *et al.*, 2002). This constancy is shared with the Abu Moharik dune field in Egypt and many other fields of barchans. Part of the story may be that the roughness of the dunes within the field slows the wind, and thus draws in sand to the field from the faster flow on either side.

Quasi-transverse patterns

Networks

Networks have also been called 'complex crescentic', 'akilé' (Monod, 1958) or 'honeycombe' dunes, as in the Kara Kum (Petrov, 1975; 40°58′N; 58°12′E; 4.5 km). The pattern develops where winds blow seasonally from three or more directions, these being obtuse bimodal and complex directional regimes (Figure 4.3d and e). The wind from each direction creates its own system of transverse dunes. Dunes related to the stronger winds survive in winds from different directions (21°37′33″N; 59°10′50″E; 200 m). The strongest or most persistent wind creates the most prominent dunes; winds from other directions modify dune slopes that had been adjusted to the dominant wind (much as with reversing dunes, but in a more complex pattern) (Figure 4.9). It is likely that more sand is shifted annually in a network than in any other dune pattern, as each wind reshapes slopes.

Reticulate dunes are a variant of networks in which the two dune-forming winds in the annual cycle are at right angles. A reticulate pattern is at 20°59′N; 58°41′E; 4.5 km in Oman.

Reversing dunes

Reversing dunes are a response to wind regimes in which two winds recurrently blow at 180° to each other (Figure 4.3f) as in some combinations of global, continental or local wind regimes (earlier). Reversal occurs daily (within a

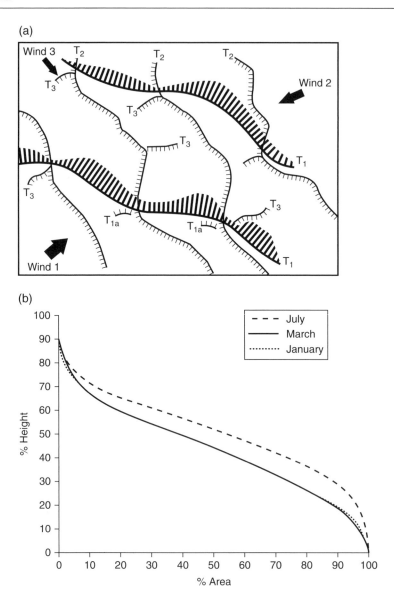

Figure 4.9 Network dunes: (a) a model of their formation; the strongest wind (Wind$_1$) creates the most prominent set of transverse ridges (T$_1$), on which are superimposed other ridges created by weaker winds from other
directions (after Warren and Kay, 1987); (b) hypsometric curves derived from repeat surveys of a network dune
system in Oman, showing the upward shift in bulk between periods of gentle winter winds and the strong south-
western monsoon (after Warren, 1988c). Hypsometric curves derived from repeat surveys of a network dune system
in Oman, showing the upward shift in bulk between the gentler winter winds and the stronger south-western
monsoon (after Warren, 1988c). Reprinted with permission from John Wiley & Sons.

particular season, such as summer near the coast); or seasonally (as with föhn-type winds, earlier). Reversing dunes created by seasonal wind reversal are revealed by surveys in the Sarykamysh Depression in Middle Asia (Figure 4.10). This pattern appears to be common in the central Asian deserts (Dubyanskiy, 1947, has some good diagrams of their dynamics). The reversing dunes in the Great Sand Dunes of Colorado (37°45'N;105°32'W; 3.5 km) respond both to katabatic winds generated on the slopes of the nearby Sangre de Cristo Mountains, blowing from one direction, and to westerlies, from the opposite direction (Andrews, 1981). The dune pattern is known locally as 'Chinese Walls'. In the Google Earth image of part of the Namib Sand Sea (24°57'S; 14°53E; 1.7 km), the dominant transverse pattern has been created by southwesterlies. A recent, north-easterly wind (perhaps a 'Berg' wind, earlier) has created reverse slip faces on almost every dune. The patterns in the Great Sand Dunes and the Namib are typical of many reversing dunes, in that the two winds seldom have the same sand-moving capability. The process of erosion by a new wind on the slip face of the earlier and opposite wind is accelerated by thresholds of movement that are lower on the high-angle slopes of the old slip face (McKenna-Neuman et al., 1997).

Star dunes ('pyramid dunes', 'dome dunes'; and vernacular names such as damkha and ghourd)

Some star dunes are low, but others are the most impressive of all free dunes, seen either from the ground or from space. They reach greater heights than any other type of dune (29°55'N; 07°53'E; 21 km). Heights of 400 m have been reported from the Ala Shan in China (39°39'N; 102°27'E; 10 km). Their only competitors in size are echo dunes (Chapter 5). They occur almost without exception in complex wind regimes (Figure 4.3e), which is, therefore, almost certainly a causal relationship. Star dunes may migrate, albeit slowly, for example where there is slight long-term or fluctuating shorter-term change in the relative velocities of the contributing winds (Wang Tao et al., 2005d).

Star dunes are formed in the same kind of complex wind-directional winds as network or reticulate dunes (earlier), but in star dunes the nodes have grown very considerably. At some point in the growth of some network dunes, they reach a size at which winds are deflected around them, after which most incoming sand is retained and feeds growth (Figure 4.11). The critical size in the Dumont dune field in California is ~100 m diameter and ~20 m height (Nielson and Kocurek, 1987).

Star dunes adopt a number of forms. Common patterns have one, two, three or more sharp-crested ridges radiate from a central peak. A large number of ridges radiating from a star dune are at 29°55'N; 07°53'E; 3 km in the Great Eastern Sand Sea in Algeria. Other star dunes are dome-like, and covered with smaller dunes with slip faces oriented in various directions (reviewed by Lancaster, 1989b). Star dunes, like linear dunes, rest on gently sloping plinths built of coarse sand, on which there may be a few, small dunes aligned to the rotating winds.

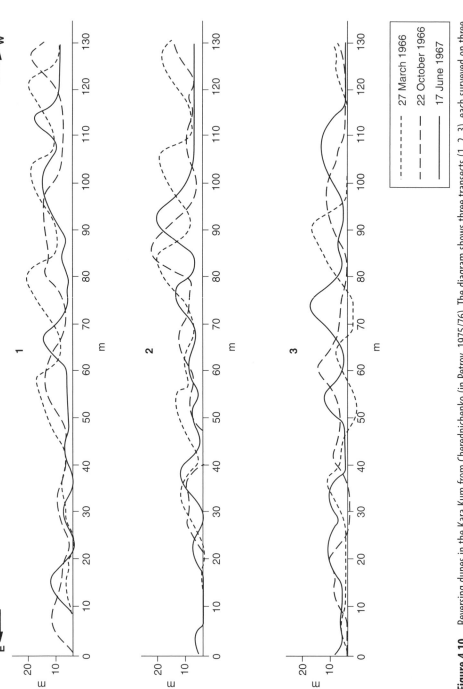

Figure 4.10 Reversing dunes in the Kara Kum from Cherednichenko (in Petrov, 1975/76). The diagram shows three transects (1, 2, 3), each surveyed on three occasions. The vertical axis (ordinate) is dune height. Steep slopes (slip-faces) face west in March; east in June.

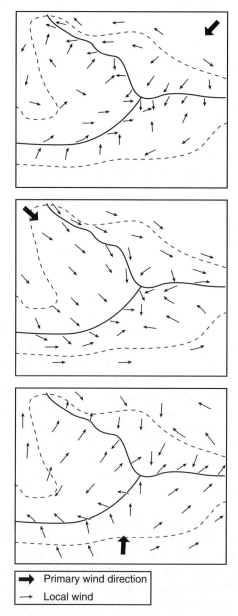

Primary wind direction
Local wind

Figure 4.11   Wind-tunnel simulation of flow over a model of an actual star dune at ~40°01′28″N; 94°48′22″E; 4 km, in north-central China (Zhang Weimin *et al.*, 2000). Permission from Elsevier, Publishers.

Two types of wind-directional regime favour star dunes. The first is the regime in mountainous terrain, where winds blowing from diverse directions are funnelled through passes and along valleys. Transverse dunes may then move up a valley, under a unidirectional wind regime, into an area of more complex winds, where they morph to star dunes (Lancaster, 1999a). The second type of wind system covers much bigger areas. It occurs where global wind systems overlap seasonally. One such overlap is between the trade winds and the westerlies, as in the Grand Eastern Sand Sea in Algeria (coordinates given earlier).

Oblique dunes

The term 'oblique dune' was introduced by Cooper (1958) to apply to some dunes on the Oregon Coast that were oblique to the resultant direction of sand movement (equivalent to the RDD, earlier). There are other oblique patterns in the Ténéré Desert in Niger, at 18°21′45″N; 13°04′13″E, 12 km. There is no good explanation for obliquity, but the Nigerien example might just be explained as an Ekman effect (earlier), over a very rough surface.

Linear Dunes

Introduction

Linear dunes are much longer and much less sinuous, and have shorter, much more symmetrical cross-sections than transverse dunes. There are linear dunes in all Earth's major sand deserts, where they cover more area than any other type of dune, and on other planets and Titan (Chapter 12). The term 'longitudinal', sometimes applied to this dune prototype, was first attached to dunes in the Great Sand Sea of Egypt, which have a north-to-south trend, which is also the direction of dominant wind. The term 'linear' is preferred here for the good reason that the Egyptian linear dunes are almost alone in being longitudinal.

There have been many classifications of linear dunes (summarised in Livingstone and Thomas, 1993). The following classification is chosen to support the subsequent argument:

Seifs are sharp-crested, elongated dunes, of various sinuosities. Most are short and presently active, as at Tsoar's classic site (shortly). A few seifs reach to 10 km, as in the Ténéré in Niger (11°00′N; 18°08′E; 19 km). Many occur on top of large linear dunes, as in the Rub' al Khali (17°39′N; 46°27′E; 12 km), in Mauritania (20°38′N; 07°33′E; 7 km), and in the Great Sand Sea of Egypt (25°18′N; 26°51′E; 7 km). No currently active seifs, as defined here, approach the length of most sand ridges (the next category), which could imply either that currently active seifs have not been active for long enough to become sand ridges, perhaps because the recent climate in now very arid areas has been more changeable than those that allowed sand ridges to develop from seifs; or that seifs have been comprehensively degraded in now semi-arid environments. Nevertheless, the following discussion shows that their separation from sand ridges is not entirely secure.

Sand ridges are much longer than seifs and occur in much larger groups, some covering up to 100,000 km². Heights are 2–30 m; widths, 150–250 m; spacing, ~200–300 m; length is commonly 20 km; some reach 200 km. Sand ridges are longer and less sinuous than seifs, and lower and less widely spaced than large linear dunes. Thus defined, they are particularly well developed in Australia (25°28′S; 136°54′E, 12 km) and in the south-western Kalahari (26°33S; 20°23′E 8 km). They are not well represented in the northern hemisphere, but there are some sand ridges with complex dendritic patterns on the Sinai–Negev border in the Levant (30°55′N; 34°21′E, 3 km; examined in more detail in Chapter 10) and others in the Orinoco basin in Venezuela, where they are mixed with parabolic dunes (06°49′N; 68°29′W; 16 km). Many, even most, sand ridges, like these, are now partially vegetated and not wholly active.

Sand ridges have many more defects than large linear dunes, 'defects' (meaning, loosely, 'loose ends', shortly). One form of defect, the Y-junction (opening upwind) is very common in the Australian and southern African sand ridges, and was remarked upon almost as soon as air photographs became available. Some sand ridge systems in the south-western Kalahari have extraordinarily complex dendritic patterns and hence very many Y-junctions; and some meander (27°08′S; 19°54′E; 5 km; Goudie, 1969). The Kalahari ridges have a much greater spread in spacing than those in Australia, and some are collected into linear quasi-wind-parallel groups that are 2.5 km wide across the wind (27°09′S; 19°53′E; 23 km). There are fewer dendritic patterns in the Australian sand ridge fields (28°56′S; 140°45′E; 13 km).

Doublet (25°28′S; 137°12′E, 20 km) and triplet sand ridges (20°46′S; 124°′E; 6 km) have only recently been reported from the Australian dune fields, in spite of their frequency (Hesse, 2010). The constituent ridges in doublets and triplets are closer together than either single dunes or the doublets and triplets themselves. Doublets and triplets are frequent in the Great Sandy Desert, suggesting that they may be associated with limited sand supply. There are more remarkable doublets in north-eastern Africa. One, on the south-western edge of the Great Sand Sea in Egypt (24°38′32″N; 25°46′00″; 14 km), is of variable width and is about 19 km long. The possible significance of doublets and triplets is explored later.

Large linear dunes are up to 150–200 m high, as in the Namib Sand Sea (23°44′S; 14°50′E; 30 km). Spacing is variable in the Namib, and in parts of the Great Sand Sea in Egypt (25°20′N; 26°44′E; 85 km). There are very much more rectilinear, regularly spaced and apparently active, large linear dunes in eastern Mauritania (El Djouf; 20°35′N; 8°47′W; 54 km; located on Figure 10.11), spaced at ~2 km; the southern Ténéré in central Niger (18°01′N; 12°39′E; 60 km; also located on Figure 10.11), spaced at ~1.2 km; and the central Rub' al Khali in Arabia

(18°34′N; 48°36′E; 100 km), spaced at ~2 km. These dunes have very few defects at their scale (later). The areas covered by large linear dunes in these three examples are huge (~120,000 km² in eastern Mauritania; ~13,000 km² in Niger and ~100,000 km² in the Rub' al Khali). Many large linear dunes are capped by seifs, whose trend usually, but not always, diverges from that of the large linear dunes on which they rest.

There are many, generally smaller sand seas of apparently active large linear dunes (defined as those having a spacing of ~2 km), few with the rectilinearity or regularity of spacing of the three large fields of the last paragraph. Some are much better studied than these Malian, Nigerien and Saudi sand seas listed in the last paragraph, and some have been extensively numerically dated (details in regional accounts of dune history later). They include the Wahiba Sands in Oman (22°22′N; 58°45′E; 4 km), the Emirates and northern Saudi Arabia (23°24′N; 54°26′E; 50 km), the central Kara Kum in Turkmenistan (40°14′N; 58°26′E; 9 km) and the central/eastern Taklamakan (38°57′N; 84°32′E; 10 km), these three last are not as regularly spaced as the African fields. The large linear dunes of the Gurbuntunggut in north-western China (44°58′N; 86°55′E; 18 km) are the most thoroughly overprinted by network dunes.

There are yet more large linear dunes with approximately the same spacing in areas that are now humid. First are those that appear to have been relatively recently stabilised, as in northern Nigeria (11°52′N; 09°55′E; 9 km; Figure 10.11), Central Sudan (13°01′N; 30°46′E; 18 km; Figure 10.10), the Pampas in Argentina (35°31′S; 61°55′W; 35 km) and south-central Kazakhstan (44°23′N; 62°49′E; 70 km). Second are apparently the much older large linear dunes in eastern Angola (15°48′S; 20°11′E; 48 km), north-central Botswana (18°44′S; 21°32′E; 90 km) and the north-western sector of the Great Western Sand Sea in Algeria (31°46′N; 00°25′E; 40 km).

This subdivision of linear dunes is conflated in much of the dune literature, implying a common process of formation and perhaps a continuum: seifs become sand ridges, given a generous supply of sand and many centuries; and sand ridges become large linear dunes, given yet more sand and yet more time. Apparent transitions from seifs to sand ridges, as at 26°35′N; 19°47′E; 70 km, in Libya may support the conflation of these two prototypes; there are fewer instances of the possible transition from sand ridges to large linear dunes.

Models of formation

Seifs

Some seif-like dunes, even some that are long, are undoubtedly extensions of the horns of asymmetric barchans (Lancaster, 1980c; 27°35′N; 29°45′E; 11 km). This may be possible only where a wind-directional regime is close to the threshold in variability that divides transverse from linear dunes (Bagnold, 1941, p. 223;

Tsoar, 1984), where there has been a recent change in wind directionality or where the underlying desert floor is sloping. Other seifs, functioning as such, are extensions of lee dunes downwind of low hills (20°14'01"N; 21°53'43"E; 6 km).

Although there have been no more than two or three field studies of active seifs, there is little disagreement about their association with bimodal wind regimes or about their mode of formation. The classic work is Tsoar's, whose site was near 30°49'N, 34°11'E; 3 km.

Three features of Tsoar's seif strongly support the bimodal-wind hypothesis. The first is its alignment with and extension in the direction of the local resultant of sand movement (the RDD, earlier; Tsoar, 1983a). The second support comes from field observations. Tsoar found a strong along-dune vortex in the lee of the dune in any oblique wind (most winds, given its alignment). He argued that this was the result of the projection, by flow over the dune, of these winds into the fast-flowing lower boundary layer, whose momentum they brought down in the lee of the dune, a process that switched seasonally from side to side of the dune. These lee winds were thus accelerated above the velocity of the oncoming wind, and were formed into vortices, as in the lee of transverse dunes (Chapter 3), which, because of the obliquity of the dune to most winds, became a dune-parallel spiral. The third feature of Tsoar's dune that endorses the 'bimodal' hypothesis became apparent after a heavy rainstorm, which penetrated deep into the dune, and the subsequent stripping by the wind of dry sand from the surface; stripping had revealed a bidirectional bedding pattern (1982; Figure 4.12), as predicted by Bagnold (1941, p. 242) and confirmed in Libya by McKee and Tibbitts (1964).

Figure 4.12   Bedding patterns revealed after rainfall on a seif dune in northern Sinai, with author for scale (photograph: Haim Tsoar). With kind permission from Haim Tsoar.

The strong along-dune component of sand flow on Tsoar's seif ensured a copious flow of sand to the downwind end, which extended the dune by 1 m in one month of one windy season, and by 32 m in two windy seasons. Most of Tsoar's findings have now been confirmed by another study of a seif, although it found no relationship between the incident angles of the wind and flow speeds and sand transport in the lee (Wang Xunming *et al.*, 2003b). A model, of the same family as those developed for barchans (earlier), has simulated the sinuous crest of seif dunes (as at Tsoar's site) and suggested that the down-dune wavelength of sinuosity is related to the seasonal alternation of formative winds (Parteli *et al.*, 2009b).

Sand ridges

In one model, which has only local application, sand ridges are the extended arms of parabolic dunes (Verstappen, 1970). It is arguable whether the model is applicable in the Thar, for which Verstappen proposed it, but some long, fairly straight dunes have, almost certainly, formed in this way (as in coastal Queensland, at 15°07′S; 145°13′E; 5 km). The model could apply only to areas that are or have been well vegetated, as near the coast in humid climates or on the temperate boundaries of the subtropical deserts (explained in Chapter 10).

There are three much more generally applicable models. The first is the received wisdom: all linear dunes, sand ridges included, are products of bidirectional wind regimes (Figure 4.3c and probably Figure 4.3d). There are two strong arguments in favour of this model. First, most linear dunes (of whatever kind) are indeed associated with bidirectional wind regimes (where known, in the present environment, and with the exceptions mentioned shortly; Breed and Grow, 1979; Wasson and Hyde, 1983b). Second, flume turntable models, cellular automata and a mathematical model similar to those developed for barchans (both earlier) show a transition between linear and star dunes (or networks), as the angle between the two main winds becomes less acute and moves towards 90°; the rate of transition depends on the relative strengths of the two winds (Parteli *et al.*, 2009b). The transition is abrupt in the turntable models, gradual in cellular automata.

The 'bimodal' model, as it applies to sand ridges, has four limitations. First, the association with bidirectional wind regimes is not universal. In the Namib, the Kalahari and Australia, there are sand ridges that are not aligned with the RDD (Lancaster, 1995a, p. 68; Bullard *et al.*, 1997; Hesse, 2010). There are two very credible explanations for these discrepancies: change of wind regime (as argued in Chapter 10 in relation to pattern-dependent dating of dunes) and inadequate wind data (very likely, given the sparse data on winds in deserts). The second weakness is the absence of a study of active sand ridges, doubtless because most are now inactive. This weakness disappears if sand ridges are seen as just another form of seif. The third weakness is that the bidirectional model, like the wind-rift model, has no explanation for the uniformity of spacing of sand ridges (or any other form of linear dune), let alone of doublets and triplets (earlier). The fourth weakness was outlined earlier: the absence of seifs (whose association with bimodal wind-directional regimes is widely accepted) in now semi-arid climatic conditions.

The second general model of sand-ridge formation has sand ridges as the outcome of a unimodal wind regime (in Fryberger and Dean's terms, Figure 4.3a and perhaps Figure 4.3b), but formed only in consolidated sediment or in association with vegetation. Dunes immobilised in these ways, it has been claimed, act as any other obstacle by creating forced lee dunes in a linear pattern (Chapter 5). The example that was used in the latest version of this hypothesis is in the Qaidam Basin in China (36°44′N; 94°24′E; 9 km), where transverse, linear dunes and yardangs parallel with the linear dunes coexist. Two samples of sediment from the linear dunes showed them to have 3–6% more silt and clay than one sample from a nearby transverse dune, which is said to imply that the linear dunes are stable enough to act as obstacles. The sediment in the linear dunes in the Qaidam may also be more cohesive owing to its higher salt content. A related model, in which vegetation plays a crucial role, is in the ascendant and is discussed further in Chapter 10, in the context of palaeodunes (which most sand ridge systems now are). This model has also been applied to the linear dunes on Titan (Chapter 12; Rubin and Hesp, 2009).

The third general model for sand ridge formation is Belknap's (1929) 'wind-rift' model: sand ridges are the consequence of the movement of sediment from interdunes to dunes, in other words of excavation, as explained in greater detail by Petrov (1975, p. 218, quoting many Soviet geomorphologists) and by Mainguet and Chemin (1983). Support for this model comes from: (1) an alluvial core in the upwind end of a sand ridge (King, 1960), and the more recent finding that some Australian sand ridges have clay-rich cores, although these are thought to be illuvial (Bristow et al., 2007b); (2) stacked palaeosols within many of the Australian linear dunes, which suggest progressive upward growth (possibly fed by excavation of the dune corridors) in arid phases and stabilisation in wetter periods (Hesse, 2011); (3) local sources of sand in some ridges (Hollands et al., 2006); (4) the scarcity in some of the Australian sand ridges of evidence of downwind extension (Hesse, 2010); (5) the absence of a discovery of increasing age downwind on any Australian sand ridges (Cohen et al., 2010); (6) near-parallel, apparently erosional lineations in other media, as in loess (in western Hungary; Sebe et al., 2011b), in hard rock (mega-yardangs as at 18°49′N; 19°24′E; 90 km, and 30°32′N; 58°13′E; 90 km; Goudie, 2007b), and in ice, where lineations are termed 'sastrugi' (Frezzotti et al., 2002).

The fourth model is the 'roll-vortex' model, which postulates that the kind of organised turbulence shown in Figure 4.13 is responsible for the formation of sand ridges (and perhaps large linear dunes, later). There is little doubt that roll vortices exist: the best (and most cited) evidence is 'cloud streets' (parallel lines of cloud aligned roughly in the wind direction). These vortices are driven by shear at the surface and buoyancy. Rotation comes from the Ekman spiral, as

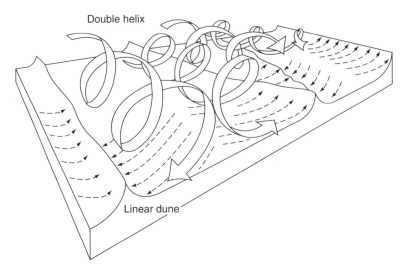

Figure 4.13   Cartoon of roll vortices, and their possible relationship to linear dunes.

modelled by Ekman (another implication of the model was explained earlier) (Young *et al.*, 2002, RL). The spacing of vortices varies from 2 to 20 km (over the ocean, the mean spacing is ~2 km), and their orientation to the geostrophic wind is up to 30° either to the left or to the right (Liu HuiZhi and Sang JianGuo, 2011, RL). The roll-vortex model has sometimes been thought to imply the wind-rift model, as by Wopfner and Twidale (2010), who rejected both (with pathos). But whereas roll vortexes might just be necessary to the wind-rift model, roll-vortices could create linear dunes without any excavation, given an upwind supply of sand.

The roll-vortex model, as will be seen shortly, could also be applicable to large linear dunes (perhaps more applicable). Earlier speculation about roll vortices, as by Bagnold, made no distinction by size, but Bagnold (1941, pp. pp. 176–180) did little more than to raise the idea that roll vortices and linear dunes might be connected. Fedorovich (1956) also played with the hypothesis in relation to linear dunes in central Asia. Folk (1976b) was the most enthusiastic advocate, although his model of vortices was strongly criticised by Leeder (1977). Tseo (1993), whose field area was in the Australian sand-ridge deserts, thoroughly reviewed the model (as also applied to subaqueous bedforms), and the evidence of its applicability to linear windblown dunes. The model might explain the remarkable parallelism of many sand ridges, and perhaps doublets and triplets (earlier). Could doublets arise where two linear dunes fortuitously developed at a spacing that favours the growth of a pair of roll vortices? Triplets are more of a problem for the imagination, and what of doublets of variable or varying width?

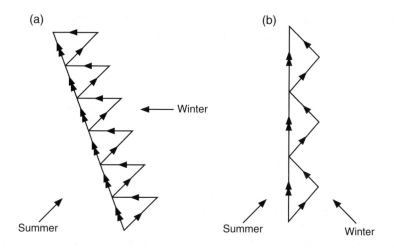

Figure 4.14   Patterns of advance of large linear dunes in (a) an obtuse bimodal wind regime and (b) an acute bimodal regime. Resultants, marked with double arrows, represent the dune extension in one annual cycle (Livingstone, 1986).

Large linear dunes

The bimodal model for the formation of large linear dunes has also been strongly supported by field measurements on a Large Linear Dune in the Namib Sand Sea (Livingstone, 2003, whose site is at ~23°34'S; 14°48'E; 9 km). Livingstone's results, however, differ in some important respects from Tsoar's model for seifs. There being no evidence in the Namib for lee-side acceleration, Livingstone argued that the seasonal flipping of wind direction was sufficient to explain the maintenance of linear form. He contrasted the acute bimodal wind regime in Sinai, which allowed rapid downwind extension, with the obtuse regime in the Namib, which allowed sand to pile up into a higher dune. This may also be one explanation for the much slower downwind extension of the Namib dunes (Figure 4.14).

The roll-vortex model (applied earlier to sand ridges) has been resuscitated in respect of large linear dunes by a model of the formation of large dunes in hot deserts, in which the size of large dunes is related to the depth of the planetary boundary layer (Andreotti et al., 2009). The extraordinarily straight and evenly spaced large linear dunes in eastern Mauritania, central Niger and the Rub' al Khali in Arabia (earlier), and the spacing of many these dunes at 2 km, conform to the patterns and spacing of cloud streets over the ocean (earlier), although these coincidences may well be fortuitous (this topic is again discussed in Chapter 10).

If Andreotti's model is to be applied to both sand ridges and large linear dunes, an explanation would be needed for the differences in their size and spacing.

If Andreotti's argument (later) about the relationship between the turbulence between dunes and the depth of the boundary layer is accepted, the question arises: at the time of their formation, was the boundary layer shallower in Australia than in the southern Sahara (and in many other localities of large linear dunes mentioned later)? Some palaeoclimatic implications of this model are discussed in Chapter 10.

Finally, there is good evidence that some of all three types of linear dune move sideways: in other words, that they are transverse dunes in disguise. Some linear dunes undeniably do move sideways. An example is the seifs in Sinai, which have moved 3 m laterally in 26 years (Rubin *et al.*, 2008); another is of the sideways movement of complex linear dunes in the Taklamakan Sand Sea (Wang Xunming *et al.*, 2004d). But there have been many periods, long or short, in which the wind regime changes subtly: sideways movement would then be possible within a longer-lasting wind-directional regime, of whatever character. Sideways movement does not necessarily invalidate most of the hypotheses, although it is probably at odds with the wind-rift model. Sideways movement had apparently little effect on the bedding pattern of the dune in Sinai (Tsoar *et al.*, 2004). The effects of an irregular alternation from strict bimodality to a more directionally variable wind regime have been modelled for linear dunes (Telfer *et al.*, 2010). The sideways movement in linear dunes has also been explored as an explanation for the putative dearth of linear-dune bedding in aeolian sandstones (Chapter 10).

Sand Sheets

Sand sheets ('sandplains' and 'coversands') are areas of gently undulating sand, of variable thickness, in which the bedding (if any) is mostly low-angle ripple strata. Some sand sheets are huge: the Selima sand sheet in north-western Sudan and south-western Egypt covers more than 120,000 km² (Maxwell and Haynes, 2001). Some of its surface is covered by zibars (later), some by chevrons (Chapter 2) and others by little more than ripples. In the Gran Desierto of north-western Mexico, sand sheets cover up to 5000 km² (Lancaster, 1995c; Figure 8.2). Smaller sand sheets are more common, as near the Great Sand Dunes National Monument in Colorado, where a sand sheet covers 710 km² (Fryberger *et al.*, 1979). There are sand sheets in the North-western Territories of Canada and in the highlands of Scotland, where they cover very small areas (Ballantyne and Morrocco, 2006; Bateman and Murton, 2006). In north-western Europe, Late Quaternary 'coversands' (essentially sand sheets) occur in formerly periglacial areas (Chapter 10). Coversands were thought to have low, chaotic patterns of relief, but laser altimetry has now revealed alignments parallel to the wind (Figure 4.15). There are further sand sheets near some coasts (Chapter 7). There are also remnants of sand sheets in ancient sediments (Biswas, 2005).

The main question about windblown sand sheets is: why is their sand not gathered into dunes? The sand sheet associated with the Great Sand Dunes in

Figure 4.15   Dune patterns on coversands near Nieuw Bergen in the Netherlands (51°37'N; 6°03'E; 8 km), revealed by laser altimetry (Jungerius and Riksen, 2010).

Colorado is ~10 m deep on average, and thus apparently capable of providing enough sand for dunes (Kocurek and Nielson, 1986). The absence of dunes could have several explanations, doubtless in various combinations:

- Gentle underlying relief, inhibiting the acceleration of winds round topography, and runoff (Kasse, 1997, and many others).
- Low rates of accumulation, as on the Bolson Sand Sheet on the New Mexico–Texas border (Hall *et al.*, 2010). Sand supply to Dutch coversands may have been inhibited by seasonal wet or frozen ground (Kasse, 1997).
- Sparse vegetation, as on sand sheets near the Great Sand Dunes in Colorado (Fryberger *et al.*, 1979). A wind-tunnel experiment, in which a

subsurface lattice of wires simulated fine roots, showed that the wires inhibited ripple formation and left a structureless deposit (Kocurek and Nielson, 1986).

- A high water table, periodic flooding or surface cementation may have inhibited dune formation (Kocurek and Nielson, 1986).
- Age. Many sand sheets are very old. Sand in the Selima sand sheet dates at 20–15,000 years, after which time it experienced a wet phase with grassland vegetation and another dry phase, when dunes could have been degraded (Stokes *et al.*, 1998b). The degradation of ancient dunes to sand sheets may have followed a change of climate in Gran Desierto, the Great Sand Dunes area in Colorado, and in north-eastern Nigeria (Kocurek and Nielson, 1986; Stokes and Horrocks, 1998). Fields of more accentuated dunes that had been degraded in a wet period might survive as sand sheets into an arid period. This is not an unlikely trajectory for some of the Saharan sand sheets, given the growing body of evidence of Holocene wet periods (Bristow *et al.*, 2009).
- Coarse, 'lag' sand accumulating at the surface may inhibit further dune formation, as perhaps in the Selima Sand Sheet, where accumulation has been slow, despite a high rate of sand flow, probably abetted for finer sands by the inelasticity of a bed of coarse sand (Chapter 1).
- Wind directionality does not seem to be a critical formative factor.

Dunes with Distinctive Sand

As explained in Chapter 1, most dunes are built of fine sand (0.10–0.40 mm diameter). 'Very fine' sands (mean diameters: 0.63 mm) are fewer but not uncommon. For example, very fine dune sands are found in much of the southern Taklamakan in western China (Wang Xunming *et al.*, 2003c), where they may be a symptom of light winds (Zu Ruiping *et al.*, 2008). The pattern of dunes in the Taklamakan, or in any other dune field with fine sand, is not, as far as is known, affected by grain size. Sands coarser than 'fine' are rarer, and some form distinctive dunes.

Gravel dunes

Gravel dunes have been reported from Antarctica (mean diameter of particles: 1.9 cm) (Bendixen and Isbell, 2007, RL) and (stretching the definition of 'dune' somewhat) in south-eastern Newfoundland (Eyles, 1976). There are gravel ripples in coastal Peru and the Antarctic (Simons and Eriksen, 1953; Ackert, 1989). The rarity of gravel dunes and ripples is a consequence of the rarity of suitable atmospheric conditions: gravel in Antarctica is moved by winds with velocities of up to $61 \, \text{m s}^{-1}$, in a dense, very cold atmosphere.

Zibars

In the Ténéré Desert in Niger (18°11'43"N; 11°04'43"E, 7 km), zibars are the low, sinuous, darker dunes between the lighter-coloured, straight seifs. Zibars are dunes built of coarse sand. They cover huge areas, perhaps 38% of the global dune field (Fryberger and Goudie, 1981). Individual fields are very extensive, as in the Majâbat al Khoubrâ of south-eastern Mauritania (Monod, 1958). Monod borrowed a lexicon of names from his cameleers for the subtly different types of zibar in the Majâbat. The most extensive is the Mréyé, which he estimated to cover 900 km² (it surrounds 19°N; 08°W; 13 km; located on Figure 10.11). Other large areas of zibars occur in the 'zibar belt' of the southern Sahara, which covers ~150,000 km² of north-eastern Niger, and >600,000 km² of northern Sudan. Zibars have been reported from many areas, even from humid-climate coastal dunes in Brazil (Barbosa and Landim-Dominguez, 2004), but most are in what are now arid or hyperarid climates.

In the Algodones Dunes of southern California, zibars reach ~2 m high and have a spacing of ~60 m; most of the Ténéré zibars are 4 m high, but some reach ~7 m, which is higher than nearby seif dunes; their spacing is 150–400 m (Warren, 1972). Other data show spacing of 60 and 100 m, and heights between 1 and 2 m (Lancaster, 1995a, p. 78). The 'mega-ripples' (not to be confused with the mega-ripples described in Chapter 2) reported from the Selima sand sheet, may be zibars in that they have similar grain size, but their spacing is far greater than those given above (20°50'N; 27°46'E; 11 km; Breed *et al.*, 1987). Chevrons, which have very low amplitude, in the same desert, are described in Chapter 2.

Zibars are effectively dome dunes (Chapter 3), in that they have no slip face. Most appear to be transverse to the resultant of winds capable of moving coarse sand. The main modal grain size in the Algodones zibars in southern California (Sweet *et al.*, 1988), and in the Ténéré, is about 1 mm. The Ténéré sediments have a second finer mode at ~0.06 mm, which is finer than the sand in the accompanying seifs. These data can be interpreted as follows: ancestral, alluvial sand with a wide range of grain size is winnowed by the wind. The medium-sized sand is moved the most frequently, given the probable distribution of wind speeds. In movement, it rebounds off the more inelastic surfaces of the coarse grains, that have accumulated on the surface (Chapter 1). The medium sand is built into seifs. When driven by rare high winds, the coarsest sands build the zibars, while the finest sand is trapped between the coarse grains and becomes the second modal size in the zibar sediment (Warren, 1971). The position of zibars on the windward side of the Algodones dunes also suggests a winnowing process (Sweet *et al.*, 1988).

By the argument that the control of the lateral (cross-wind) dimension of transverse dunes is grain size (Hersen, 2004; earlier), zibars should have wider arcs than dunes of finer sand, and this does seem to be the case. The coarseness of their sands may also explain the parabolic shape of many zibars (curves opening upwind), as in south-eastern Libya (21°15′N; 21°21′E; 7 km). Rough beds of coarse sand induce a velocity gradient that is steeper than that over smooth fine-grained beds, so that, when coarse sand on the crest of a zibar (and therefore the crest itself) is in movement, its base is immobile. Zibar crests might therefore move forward of the 'horns', creating a 'parabolic' shape (given a certain distribution of strong and gentle winds).

Clay dunes

Clay dunes were first reported from the Gulf coast of Texas by Coffey, who imme-diately realised that they had been built of clay-rich pellets (1909; Chapter 1). Many more clay dunes have since been identified, some with clay contents of 77%. Apart from pellets (Chapter 1), some clay is added to clay dunes as flakes (Dare-Edwards, 1982).

The cross-sectional shape of most clay dunes differs from that of sand dunes, in that slopes are gentler than in sand dunes and may be steeper to the windward side than in the lee. Many clay dunes, like some in the Mojave, are wind-eroded into yardangs (Blackwelder, 1934) or, because their salt content discourages veg-etation, deeply gullied by runoff. Clay dunes can form only where there is a rainy season during which infiltrating rainwater disintegrates the pellets in the dunes and welds it into a coherent mass (Dare-Edwards, 1982).

Lunettes

There are good examples of lunettes at 28°58′44″S; 25°37′08″E; 4 km in the Orange Free State. Many, but not all, lunettes are clay dunes, and many, but not all, clay dunes are lunettes. Lunettes commonly reach 5 m high, some many tens of metres (Holmes et al., 2008).

A model of the growth of a lunette starts with a lake that is shallow enough to allow both waves to disturb and entrain from the bed; and circulation within the lake that is more horizontal than vertical. It is this circulation that carries sus-pended sediment towards the lee shore, where it forms a bar, which itself then becomes the base of a lunette. A dry season, or an extended drought, is necessary for the formation of a lunette, because the lake bed is then excavated by wind erosion, which both perpetuates the lake and yields material for the dune (Bowler, 1973; Lees and Cook, 1991). Some lunettes, like clay dunes, have formed in humid environments, as near the coast (as in Coffey's in Texas). It is wave action that creates the curved shoreline, not any windblown process (except in driving the lake's circulation and the waves that lift particles off the lake bed).

Many lunettes are sandy, either in layers within a clay dune, as arcs of a full lunette, or even as whole lunettes (Holmes et al., 2008). Sandy lunette dunes behave like other sand dunes, for example in being mobile and having their steepest slopes on the downwind side. Not all lake-margin dunes are lunate, despite being termed 'lunettes' (because of other shared characteristics). Many 'lunettes' in Tunisia, for example, are not lunate, but share most of the characteristics and modes of formation of lunettes (Coque and Jauzein, 1967).

Gypsum dunes

Gypsum dissolves and is leached from the soil in wet environments, but in dry enough conditions, evaporation of water rich in calcium sulphate produces sand-sized particles of gypsum that survive to behave like quartz sand, although they are more liable to be reduced to dust during saltation. Gypsum dunes are not common but cover some big individual areas. The biggest is probably White Sands in New Mexico (~850 km^2; Kocurek et al., 2007; 32°48'N; 106°17'W; 1.5 km). There are others at Lake Amadeus in central Australia, in southern Tunisia (33°50'N; 8°36'E; 3 km) and on Mars. The forms of gypsum dunes differ little from those of quartz sand: at White Sands, there are barchans, transverse dunes and parabolic dunes. There are distinctive long-term rhythms in the formation of gypsum sands (Chapter 10).

Diatomite sands

The deposits of freshwater palaeolakes are characteristic of a broad latitudinal swathe of the southern Sahara (Gasse, 2000, RL). The exo-skeletons of diatoms in these lakes accumulate as grey sediment with a very low specific gravity, which forms weakly cohesive flakes when it dries, and these flakes are easily raised by the wind, in which they break down rapidly. In some places, sand-size pellets survive for long enough to be built into dunes. In the Bodélé Depression of northern Chad, where winds are fierce, many of these dunes are ~30 m high (Warren et al., 2007), and their celerity has been found to be faster than any others on the record (16°49'N; 16°55'E; 25 km; Chapter 3). However, these sands are quick to turn to dust in saltation and in slip-face avalanches. These fragments are the source of the dusty Harmattan wind that blows westward across West Africa, and reaches Peru. The fragility of the sands also ensures that very few of these dunes survive for more than 5 km from the western edge of the lake deposit that is the source of their sediment (as in the quotation for last Google Earth scene).

Volcanic sands

Much of the dune sand in Iceland, and in some smaller areas, is heavy and dark, being of volcanic origin (Mangold et al., 2011). Even more rarely, dunes are created in volcanic debris by the high winds from the volcanoes themselves. In these, dune spacing decreases away from the vent, as the winds slow (Fisher and Schmincke, 1994). A very high proportion of Martian dune sand is volcanic in origin (Chapter 12).

Snow and ice dunes

Dunes built wholly of non-coherent, dry, deeply frozen snow or ice particles are common, even occurring occasionally, if fleetingly, in England. They are more permanent and take on a number of forms where thaw is rarer (Sturm and Liston, 2003), including linear dunes in bidirectional wind regimes, as in Ellesmere Land in the Canadian Arctic (Lewkowicz, 1998), and in the Antarctic (Godwin, 1990). Dry snow, like sand, also builds forced dunes, including nebkhas (Chapter 6).

Active 'mega-dunes' built of ice particles cover 500 000 km² on the East Antarctic Plateau, where they are driven by strong katabatic winds. They adopt transverse or reticulate patterns with spacings of the order of 2 km. They have a much lower amplitude (~2 to ~10 m) than quartz-sand mega-dunes; no slip faces; very coarse grain sizes (0.5–1.5 cm); and a great range in grain size. They may form in a fundamentally different way: standing atmospheric waves have been favoured as an explanation (Anschutz et al., 2006).

Niveo-aeolian deposits

Mixtures of snow and sand do not form distinctive dunes so much as distinctive dune deposits. They are the result of the alternating deposition of snow and sand. In some years, and on some dunes on the eastern coast of Hudson Bay in Quebec, niveo-aeolian deposition accounted for 75% of all dune deposition (Bélanger and Filion, 1991).

Interest in niveo-aeolian deposits has been driven largely by attempts to interpret periglacial deposits in what are now temperate areas. In cold environments, there have been reports of collapsed surfaces and structures and hummocky terrain, attributed to the thawing and re-freezing of deposits of alternating snow and sand (Bélanger and Filion, 1991), but few of these structures have been found in palaeodunes. Indeed, experimental work with layers of snow and sand did not produce this kind of structure, although it did produce distinctive 'spongy', porous textures (Dijkmans and Mücher, 1989).

References

Bendixen, Q.D. and Isbell, J.L. (2007) 'Gravel dunes formed by aeolian processes in a cold, dry environment, Allan Hills, Antarctica: a possible Mars analogue', *Geological Society of America, Abstracts with Programs* 39 (6): 205.

Burt, F.A. (1949) 'Origins of geologic terms', *The Scientific Monthly* 69 (1): 20–22.

Gasse, F. (2002) 'Diatom-inferred salinity and carbonate oxygen isotopes in Holocene waterbodies of the western Sahara and Sahel (Africa). *Quaternary Science Reviews* 21 (7): 737–767.

Halevy, G. and Steinberger, E.H. (1974) 'Inland penetration of the summer inversion from the Mediterranean Coast in Israel', *Israel Journal of Earth Science* 23 (1–2): 47–54.

Liu HuiZhi and Sang JianGuo (2011) 'Numerical simulation of roll vortices in the convective boundary layer', *Advances in Atmospheric Sciences* 28 (3): 477–482.

Markowski, P.M. and Richardson, Y.P. (2010) *Mesoscale Meteorology in Midlatitudes*. Chichester, UK: John Wiley and Sons.

Young, G.S., Kristovich, D.A.R., Hjelmfelt, M.R. and Foster, R.C. (2002) 'Rolls, streets, waves, and more – A review of quasi-two-dimensional structures in the atmospheric boundary layer', *Bulletin of the American Meteorological Society* 83 (7): 997–1001.

Chapter Five
Forced Dunes

Dunes Built around Bluff Obstacles

'Forced' dunes are those built around obstacles that have steep slopes facing the wind, like bushes, boulders or hills ('bluff bodies') (the distinction between 'free' and forced' dunes was explained in Chapter 3). The size of a forced dune is related to the size of the obstacle; its form is independent of size. Small forced dunes do not have the capacity to withstand changes of wind direction and are therefore frequently reoriented, usually seasonally (as in Figure 6.4). Big anchored dunes, as in Peru (9°22′25″S; 78°20′24W″; 6 km), are better buffered, so that only their surface is modified (usually also on a seasonal basis), while their major features survive from year to year. Figure 5.1 gives a nomenclature for forced dunes.

Climbing and echo dunes

A site that illustrates many types of these dunes is on the north-eastern margins of the Aïr Mountains in northern Niger, between 19°53′N; 08°41′E; and 20°11′N; 08°40′E, all at 19 km eye altitude (located in Figure 10.11). Wind-tunnel studies explain some of the characteristics of these dunes. If the windward slope of the obstacle is less than ~30°, sand is transported up and over it. On a slope between 30° and 50°, sand is deposited in a climbing dune on the upwind slope. Climbing dunes grow to a stable configuration, over which incoming sand is then transported (there is a range of these slopes and dunes around 19°55′14″N; 08°43′36″E; 5.5 km; Tsoar, 1983b; Qian Guangqiang et al., 2011). Most climbing dunes accumulate in

Dunes: Dynamics, Morphology, History, First Edition. Andrew Warren.
© 2013 John Wiley & Sons, Ltd. Published 2013 by John Wiley & Sons, Ltd.

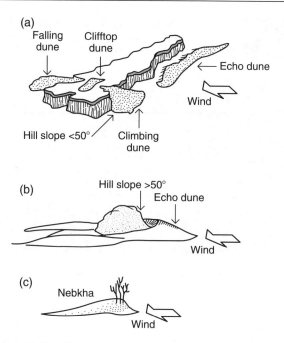

Figure 5.1 Nomenclature for forced dunes.

windward gullies, where, if the gully sides are steep, the windblown sand may be interleaved with rockfalls and coarse alluvium (Turner and Makhlouf, 2002). Where a climbing dune abuts a rock slope, a gully, taking occasional runoff, may be incised at the boundary between sand and rock (as in California at 34°04′57″N; 115°38′16W; 940 m; further illustrations in Zimbelman *et al.*, 1995).

On windward slopes steeper than ~50° (Tsoar), or ~60° (Qian GuangQiang and colleagues), the approaching wind forms a vortex upwind of the scarp, and this creates return flow (counter to the approaching wind) on the surface. The turbulent vortex sweeps out a sand-free area upwind of the abrupt slope. Where the vortex loses strength, an 'echo dune' is formed, upwind of the hill. The growth of this dune is augmented by sand brought by the approaching wind, which allows some echo dunes to grow to great heights, as on the image of the western slope of the Aïr Mountains (coordinates given in the last paragraph).

Flanking and lee dunes

If the bluff obstacle has a low flank, the frontal eddy escapes towards it, and creates a flanking vortex. If the bluff obstacle is narrow across the wind, the frontal vortex splits into two flanking vortices (as at 20°18′02″N; 09°00′56″E; 2.50 km). Flanking

Figure 5.2   Echo dune upwind of a large nebkha (defined in Chapter 6).

vortices, like the upwind vortex, sweep out corridors between the fixed obstacle and dunes on their flanks. The flanking dunes, like those formed by the frontal vortex may also intercept sand brought from upwind (Figures 5.1 and 5.2).

In the lee of a bluff obstacle, turbulence is enhanced, as in the lee of the Amboy volcanic crater in the Mojave Desert (34°32′41″N; 115°47′20″W; 4 km; Greeley and Iversen, 1987). At Amboy, turbulence sweeps sand away from the sand-free shadow in the lee of the cone, whose darker colour contrasts with the lighter colour of the rest of the lava field, on which there is a dusting of windblown sediment. At a wide range of Reynolds numbers, the turbulence in the lee of obstacles like this takes the form of counter-rotating 'Kármán' vortices (Markowski and Richardson, 2010, p. 354, RL). The enhanced turbulence sweeps sand from a 'shadow' zone.

In some cases, the sand-free shadow in the lee of the parent obstacle narrows downwind, as at Lake Faguibine in Mali, which extends >100 km downwind and covers ~540 km^2 (16°46′N; 03°50′W; 80 km; located on Figure 10.11). At Faguibine, the lee hollow formed downwind of a number of bluff sandstone plateaux (to the ENE), when the surrounding dune field was active sometime in the Late Pleistocene, when winds may have been stronger than at present. In other cases, the flanking dunes run parallel to each other for a considerable distance downwind, as in the lee of the Amboy crater, and the many small plateaux in the area of 22°17′N; 11°57′E; 100 km. This difference in behaviour may be a function of height, or of a bimodally directional wind regime, which could explain the tapering lee hollow at Faguibine. An even lower, and narrower plateau, at 20°14′01″N; 21°53′43″E, is the origin of a single seif-like dune trailing downwind for many kilometres. Single lee dunes also develop downwind of plants (Chapter 6).

Bagnold's account of dunes formed around bluff obstacles is short (1941, pp. 189–195) but adds two further categories of forced dune to the list in this chapter. The first is dunes formed downwind of a narrow gap between two obstacles, in which flow is constricted and accelerated (the 'Venturi' effect). Downwind of the gap, the flow expands and loses velocity. If there is sand in the wind, little of it forms dunes until the velocity has fallen sufficiently, as at 23°16′26″N; 12°09′46″E; 35 km, where a large dune has formed ~5 km downwind (south-west) of a gap in an escarpment. The second category is of dunes in the lee of a serrated escarpment, as north of the lakes at Wanyanga Kebir in northern Chad, at 18°56′27″N; 20°53′00″E 9 km, located on Figure 10.11.

Cliff-top and falling dunes

In the reduced wind velocity just beyond the crest of the 30–50° category of windward slope, cliff-top dunes may be built with the sand that has been carried up sand ramps. Cliff-top dunes have accumulated on the summits of some sea and lake cliffs (as at Rubjerg Knude in northern Denmark; 57°26′53″N; 9°46′25″E, 550 m view in 3-D from west; Saye *et al.*, 2006). If there are no cliff-top dunes, or if sand escapes from them, it may be taken to falling dunes on the lee slope. Most falling dunes, like climbing dunes, accumulate in valleys or clefts (23°55′N; 11°34′E; 800 m). Where winds flow over abrupt leeward cliffs, as in some coastal sites, flow is reversed, much as in the lee of a dune. This flow can build up a dune against the cliff, rather as an apron builds against the lee of some slip faces (Chapter 3; Hesp, 2005).

Dunes on Gently Sloping Terrain

The effects of gentle slopes on dune patterns have been analysed only recently. A wind-tunnel study of a wind blowing more or less at right angles over a 'valley' with gently sloping sides found that the wind was accelerated in a zone at least 1.5 valley widths upwind, and two valley widths downwind. Dunes on the downwind side of the valley may be deflected from their prevalley orientation, or adopt a simpler pattern (Garvey *et al.*, 2005; 26°28′S; 20°36′E; 6 km). When the wind approaches the valley obliquely, it creates along-valley flow, in which dunes are aligned with the valley.

Reference

Markowski, P.M. and Richardson, Y.P. (2010) *Mesoscale Meteorology in Midlatitudes.* Chichester, UK: John Wiley and Sons.

Chapter Six
Dunes and Plants

This chapter is about active dunes. Dunes wholly stabilised by vegetation are described in Chapter 10. Most of the dunes examined here are features of the wide border zone between deserts, where almost all dunes are mobile, and areas where plant cover is complete enough totally to prevent the movement of sand. These border areas and this chapter share many dune forms with coastal dunes, where sand is supplied at a rate that competes with the growth of vegetation. Only the exclusively coastal types of dunes are discussed in Chapter 7.

Wind, Sand and Plants

If a dune is to be built around a plant, sand must blow, and the plant must grow. The interaction of these two processes, which determines the form of the resulting dunes, can be understood at six levels of complexity.

Rigid objects

The first and simplest is to represent plants and their substrate as rigid, and the plants as very simple shapes, in order to see how they share or 'partition' the drag of the wind. Studies in wind tunnels (Figure 6.1) have shown that partitioning (λ), defined as

$$\lambda = nwh,$$

Dunes: Dynamics, Morphology, History, First Edition. Andrew Warren.
© 2013 John Wiley & Sons, Ltd. Published 2013 by John Wiley & Sons, Ltd.

Figure 6.1   Guelph University wind tunnel ready for a study of the effect of one particular arrangement of rigid obstacles on the sharing of drag between the bed and the obstacles (Brown *et al.*, 2008).

where n is the number of obstacles per unit area; w is their width; and h, their height, determines the degree of partitioning.

Experiments like this have shown that the shape of the rigid obstacles makes little difference to partitioning (Raupach, 1992). They have also shown that an obstacle protects a 'wake', which is a triangular area narrowing downwind from the lee side of the obstacle 7.2 comparable with the shapes of some protected areas in the lee of much larger obstacles (Chapter 4), although some models define the protected area as straight-sided protrusion, which fades downwind (Okin, 2008). When λ is small, the wake extends downwind to ~10 times the height of the plant; field measurements in Burkina Faso found that the protected area was about 7.5 times the height of the plant, this lower value probably being a result of fluctuation in the direction of the wind (Leenders *et al.*, 2011). The ground beneath the wake, like the ground covered by the plant itself, is protected from erosion.

Spatial pattern

The second level of complexity in plant/sand interactions is the effect of the spatial pattern of the rigid obstacles. In general, if the density of obstacles is held constant, the pattern has little effect on partitioning (Brown *et al.*, 2008), but when the obstacles are close enough together, the wind is accelerated between them (Ash and Wasson, 1983, quoting Garratt, 1977). On a loose sandy surface,

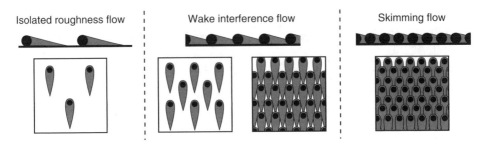

Figure 6.2   Interaction of winds with plants at increasing cover (Wolfe and Nickling, 1993). Permission from: Sage, Publishers.

this acceleration can cause erosion, which, in turn, can threaten the survival of the plants on either side of the gap. The effect should help to keep the plants at a distance from each other (another, probably more pervasive, process keeping plants apart is discussed later in relation to nebkhas). However, a progressive increase in the number of obstacles per unit area decreases and eventually eliminates shear on the bed, and thus sand movement (Figure 6.2).

Porosity

Another layer of complexity is added by examining the effect of the porosity of the obstacles. This is also discussed in reference to the effect of the porosity of sand fences on their ability to slow the wind and to trap sand (Chapter 13). In the field, the effects of porosity are seen in a comparison between two species of shrub in the arid parts of the western United States: greasewood bushes (*Sarcobatus* spp.) collect more sand than creosote bushes (*Larea tridentata*), whose foliage is more porous (Gillies *et al.*, 2000). When there is little porosity, wind and sand are deflected round the plant; when there is more porosity, the plant provides less shelter and allows more wind and, if present, sand to pass through the gaps. Litter is much less flexible than living plant material and can also restrict sand movement but can be treated as a rigid set of obstacles (Chapter 5). The coverage and cover pattern of litter are as variable as that of the plants that shed it (Wiggs *et al.*, 1995a). Litter is especially persistent in dry climates, where the fungi and bacteria that might rot it are less active than in humid climates (Murphy *et al.*, 1998, RL).

Flexibility

Flexibility is another relevant property of living plants. Extreme bending reduces the shelter afforded by the plant, by reducing its effective size, and swaying in high winds increases turbulence, which may create high, short-lived velocities capable of keeping sand in the wind (Gillies *et al.*, 2002; Udo and Takewara, 2007; Field *et al.*, 2012). But bending may also protect the surface, if it absorbs energy and slows the wind near the ground, and this suggests that grasses, which are

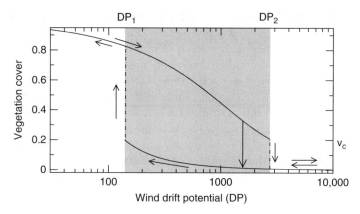

Figure 6.3   Two-parameter plot of outputs of the mathematical model of Yizhaq *et al.* (2007). Explanation in text. Reprinted figure with permission from Yizhaq *et al.* (2007). © 2007 The American Physical Society.

more flexible than woody plants, might provide greater protection (hence the replacement of grass by mesquite might increase wind erosion, shortly).

Plants as living things

But plants are alive. Their survival is endangered by sand-blast or burial in sand, and they have problems of establishment on loose substrates. Interactions at this third level of complexity have been studied in the field and by mathematical modelling. Many of the field studies have been undertaken in the semi-arid plains of the south-western USA, where mesquites (a number of species in the *Prosopis* genus), have been spreading over the last century and a half, perhaps, in part, because of the introduction of cattle. The spread of mesquite has been at the expense of grassland, which, as explained earlier, might have better protected the soil from wind erosion.

These studies have shown that where winds come dominantly from one direction, some of the gaps between mesquite plants are elongated to become 'streets'. Elongation is driven by sand blowing towards the upwind end of a nascent street, where it may destabilise the seed bed, may bury emerging plants and may sand-blast any plants that do manage to survive the other hazards. In one study, streets were 50–100 m long and up to 5 m wide. Their size and distribution were strongly related to the wind-driven mobility of the substrate, and hence were greatest where the substrate was sandy (Okin *et al.*, 2006). In these areas, it is the streets that provide the sand for nebkhas (shortly). Streets also supply most of the dust that is blown from these areas.

A mathematical model at this fourth level of complexity brings together the effects of the (sand) drift potential (DP) (Chapter 4); the celerity of dunes (Chapter 3); the growth rate of plants; and the effects of the exposure of roots, burial of seedlings and sand-blast. One output of this model is shown in Figure 6.3, on which the upper line traces the coverage of plants in response to increasing DP. The thinning of the plant cover is at first gradual, but cover then declines precipitously towards bare ground. If, subsequently, the DP declines (the wind climate moderates), the lower line shows that recovery is at first slow but then suddenly accelerates. The decline and recovery paths are thus quite different and define a range of critical cover values (the grey box in Figure 6.3); in short, there is hysteresis (Yizhaq *et al.*, 2007). This bistable relationship between plant cover and sand mobility may explain the finding that there is no single cover value that separates active from stabilised dunes (Hesse and Simpson, 2006, and shortly). It may also explain some of the instability of the border region between active dunes and fully stabilised areas, although instability is also caused by external changes, as seen shortly. The bistable relationship may further explain the many sharp boundaries between moving sands and almost fully vegetated areas (as at 26°34′N; 69°35′E; 6 km in south-eastern Pakistan).

The broader time/space framework

The sixth and final level of complexity coarsens the focus to areas of a few tens of square kilometres and to lengths of time of the order of decades. Studies at this scale are reality checks. At this scale, patterns both of mobile sand and of vegetation are very dynamic, under the influence of at least four controls: (1) high spatial and interannual variability of rainfall (and therefore in plant cover); showers can saturate a small area, while hundreds of metres away, the drought continues; seasonal rhythms are changeable, and droughts may last from months to decades; (2) the dispersal and concentration of runoff at many scales, from small puddles to flood plains and ephemeral lakes; (3) temporal fluctuations or spatial patterns in the depth of the water table, which may be controlled by factors other than rainfall; (4) the variation in the water delivery of soils to plants, depending on infiltration, soil texture, arrangement of soil horizons, etc.; sandy soils are generally better vegetated in dry areas, mainly because of their high infiltration, and low capillarity (controlling water loss by from evaporation) (Tsoar, 1990c); another relevant property of sandy soils is discussed in Chapter 14 in the context of the differential response of arid lands to disturbance; (5) lag effects: high rainfall or severe droughts increase or reduce the cover of plants and so withhold or release sand for dune building, effects that may only be apparent and persist long after the event (Bullard *et al.*, 1997); and last, (6) secular changes and spatial patterns in wind power. These processes combine to ensure that there are few clear-cut patterns of cover/bare sand, as in the Australian sand-ridge deserts (Hesse and Simpson, 2006).

Dunes among Plants

Nebkhas

Nebkhas are dunes collected around plants (Figure 6.4).

Nebkhas are features of the borderland between the desert and the sown; humid-climate coastal dunes (Chapter 7); and windy hilltops in humid or cool parts of the world, as in north-western Ireland (Wilson, 1989). The best

Figure 6.4   Small nebkhas with single lee dunes.

nebkha-forming plants are those that can maintain growth as sand accumulates, as well as being hardy in the face of drought, salinity, sand blasting, or extremely low temperatures, as in places like central Asia. *Tamarix* spp. (Figure 5.2), *Salvadora persica* (26°05′03″N; 69°18′37″E; 300 m), mesquite (earlier) and a limited number of other species are common nebkha formers. *Haloxylon* spp. (Saxauls) are common nebkha-gathering plants in central Asia. On coastal dunes, *Ammophila arenaria* (marram or oyat) is the best known of these blown-sand-tolerant plants (Chapter 7).

The constituents of nebkhas reflect the nature of the ambient windblown sediment: they range in size from silt to coarse sand, and from saline to sweet. Clay may come as flakes or pellets, as to clay dunes (Chapter 4) or dust (Nickling and Wolfe, 1994; Khalaf *et al.*, 1995). The rate of growth of nebkhas probably covers a huge range, depending on variations between species and the same species at different stages in its life cycle, the recent history of rainfall, the spatial and temporal variation of local sediment supply and so on.

Short lee dunes develop downwind of nebkhas in environments that are sandy and in which vegetation is sparse enough to allow the wind to move the sand, as suggested by wind-tunnel studies (Figure 6.2; and 27°58′06″N; 12°46′44″W; 225 m). Many of these 'shadow dunes' have distinctive bimodal bedding, similar, but at a smaller scale, to the bedding patterns of linear dunes, both being formed by the interplay of winds from different directions, albeit at different scales (Gunatilaka and Mwango, 1989). Bedding in the cores of most nebkhas is destroyed by bioturbation (by roots, burrows and so on), but the cores of *Tamarix* nebkhas accumulate in stacked concentric cones of alternating litter and sand, the depth of each stratum varying with the production of litter (more in dry years), and the addition of new sand (more in windy years). The saline litter from the *Tamarix* discourages bioturbation (Qong Muhtar *et al.*, 2002).

Nebkhas do not form everywhere in semi-arid country. In different parts of a dune landscape, and different climatic zones, there are different outcomes of the competition between the erosion that might eliminate a nebkha, and the supply of sand, and of moisture that may ensure the survival of the parent plant. On a semi-arid site in coastal Israel (mean annual rainfall ~200 mm), the fastest growing nebkhas are at the crest of a moving dune, where they are maintained by a high rate in sand supply. However, the onward movement of the dune eventually isolates these nebkhas on eroded, shrinking pedestals, which hold progressively less moisture, leaving plants to wither and die, depriving their nebkhas of an anchor, and allowing them to be destroyed by the wind (Ardon *et al.*, 2009). This 'pedestalisation' may also be a consequence of wind erosion in the accelerated flow between nebkhas, if they are appropriately spaced (earlier; Wang XunMing *et al.*, 2006b). In south-western New Mexico, where the supply of sediment is limited, nebkhas in upwind positions capture sediment that might have reached downwind. The consequence is that nebkhas become smaller downwind of deflation hollows (Langford, 2000).

The first step in understanding nebkhas is to distinguish them from mounds formed by non-aeolian processes, such as rainsplash, runoff and even seismic activity (reviewed in Cooke *et al.*, 1993, pp. 356–357). Windblown and the other processes combine in diverse ways to build mounds, and it is not always easy to detect their proportional contribution. The windblown sediment in nebkhas is most commonly supplemented by sediment moved in by rainsplash, which is more intense outside than within the canopy, and by sediment brought by runoff (Buis *et al.*, 2010).

The second step is to see nebkhas as systems in which the growth of the plant and the supply of sediment are independent controls. Within Yizahq's two thresholds (earlier), the life cycle of the plant is the main control of the size and survival of the dune. Nebkhas also reflect the sizes and branching characteristics of their parent plants. Even within one species, a comparison between nebkhas showed a strong relationship between canopy volume and the nebkha volume (Zhang Pujin *et al.*, 2011). Where there are different plant species within the same area, there are larger nebkhas beneath plants with closed canopies than beneath prostrate or upright species (Hesp and McLachlan, 2000). Nevertheless, nebkhas in any one area are usually of a similar size, both because few species have the shapes that favour the accumulation of sand and because of the low plant diversity in most of the dry and coastal world (Cowling *et al.*, 2004, p. 153, RL). Nebkhas also have a plant-dependent demography. In central Tunisia, growth rings on stems showed that the ages of nebkhas were 50–200 years, following the cycle of growth, maturity and decay of the parent plants (Tengberg and Chen Deliang, 1998). In many areas, nebkhas are evenly spaced, which could be for one or both of two reasons. The first was explained earlier: an increase in wind erosion between plants that are too close together. The second reason is more general. It bundles together many processes that might create equal spacing in plants in semi-arid areas, which is common on both sandy and other substrates, and which has puzzled ecologists for decades. The balance of opinion appears to be that equidistance is largely the result of competition between plants for water (Kambatuku *et al.*, 2011, RL). If the groundwater is saline, the growth of a nebkha can raise the rooting depth of plants sufficiently far above the saline groundwater to allow faster growth of the parent plant, and with it the nebkha (He XingDong *et al.*, 2003).

But, as with the general controls on the relationship between plants and sand movement (earlier), there is a third and vital step to understanding nebkhas. It is to acknowledge the positive feedback between growing plant and accumulating sand. New sand and dust bring nutrients and enhance the water-holding capacity (Dougill and Thomas, 2001), and as the plant grows, it traps an increasingly higher proportion of the available sand and dust than the surrounding country (Field *et al.*, 2012). The new sediment allows the plant to extend its roots and to grow more vigorously (Bendali *et al.*, 1990). There is then positive feedback as the nebkha grows with the plant. The increasing stability created by these two processes may allow colonisation by lower-growing, more disturbance-averse annual/ephemeral plants, lichens and microflora, and by burrowing vertebrates and invertebrates, all of which bring in more nutrients and organic matter, which in turn binds, stabilises and further encourages the growth of the nebkha (Yue XingLing *et al.*, 2005). Some species of nebkha-formers produce litter that is more attractive to these colonists than others (Hesp and McLachlan, 2000). A model of nebkha growth, using Werner's cellular automaton (Chapter 4) with a vegetation component, follows this logic, by allowing the growth of the plant to be stimulated when sand is supplied to it (Baas and Nield, 2010).

Two final observations: first, small as they are, some nebkhas are relics. OSL dating (Chapter 10) has shown that nebkhas at many sites in the Arkansas River valley, Ozark Plateau and coastal plain of the Gulf of Mexico date to a period in the Holocene known to have been dry (Seifert *et al.*, 2009). Second, there are claims that nebkhas are symptoms of degradation, because of a change in either climate or land use, although this is contested (Dougill and Thomas 2001). Some may be, but nebkhas are ambiguous symptoms.

Blowouts

Blowouts are in Belknap's (1928) 'wind-rift' category, in that they are created by the erosion rather than the construction by the wind. There are blowouts at 52°08'51"N; 4°20'37"E; 1.5 km in the Netherlands (where they are encouraged in order to maintain the full range of ecological niches in coastal dunes; Chapter 14). A blowout begins with a breach in plant cover over a critically sized area, whose dimensions depend on the morphology and distribution of plants, position on the parent dune, ambient wind speeds and rainfall (which encourages the growth of the gap by inducing runoff erosion, or discourages it by enabling vegetation to establish and grow). The initial breach may be created by: wave attack on the fore-dune; paths; burrowing animals (such as rabbits; Rutin, 1992); heavy fallout of sand entrained upwind; desiccation of upper slopes during hot, dry summers; extreme winds; gusty winds (Jungerius, 1984; Mangan *et al.*, 2004); gullying after intense rainfall; and such like (Anderson and Walker, 2006); or, commonly, by two or more of these processes. The longer axis of a blowout is commonly parallel to the direction of the prevailing wind.

Blowouts cover variable proportions of dune fields. Eighteen per cent of dunes on the shores of Lake Huron in Canada were covered by blowouts in 1998 (Dech *et al.*, 2005). On the Manawatu coast in New Zealand, blowouts occur at 20 per kilometre of fore-dune (Hesp, 2002b). The size of a blowout varies from place to place (as on the Manawatu coast), and from time to time. The modal length of blowouts on the Meijendel dunes in the Netherlands was 25–30 m in the late 1980s, whereas on the De Blink dunes in the same years, the modal length was 15–20 m (Jungerius and Schooberbeek, 1992). In north-western Ireland, some blowouts are 45 m deep (Carter and Wilson, 1993). Two subtypes have been recognised, saucer (broader) and trough (narrower), which exist side by side on many dunes (Hesp, 2002b).

If the wind entering the blowout comes over a steep face, there is flow separation and recirculation, as in the lee of a free dune (Chapter 3). These and other winds accelerate through the blowout, and if it is narrow, they may reach velocities well above that of the ambient wind. Patterns of flow may be complex: accelerated 'jets' and vortices may occur at various heights (Hesp and Hyde, 1996). In some blowouts, winds from different quarters may cooperate to move, deposit and ultimately remove sand (Hugenholtz and Wolfe, 2005b). In one blowout, high winds had less effect on erosion than moderate but more frequent winds (Jungerius *et al.*, 1991). In other situations, high winds fill rather than empty a blowout (Jungerius *et al.*, 1981).

The walls of blowouts are usually loosely bound by plant roots, and in many cases further cohesion is given by moisture, both of which characteristics help to maintain slopes that are steeper than the angle of repose of loose sand (Chapter 3). Tree roots may help the sand to bind even 5 m beneath the surface (Wortham, 1913), but the roots of most dune plants are shallower and, as such, support steep slopes only on the upper slopes, below which there may be slip-faces in loose sand. The sides of blowouts recede partly by the removal by the wind of sand that cascades down the over-steepened lower slopes, and partly by the undermining of blocks of sand held together by roots. Where it has been measured, erosion is most intense on the base of the bowl, but if it uncovers a shallow water table or a pebbly or cohesive stratum, erosion is halted (Fraser *et al.*, 1998). Enlargement usually occurs in seasons when plant cover is reduced, as in the dry seasons of monsoon climates, in the summers in temperate regions or in severe winters, as in eastern Canada (Pluis, 1992; Byrne, 1997). In Prairie Canada, south-facing slopes dry more quickly and erode more readily (Hugenholtz and Wolfe, 2006).

The wind decelerates where a blowout widens downwind, and sand may then be deposited. In some places, a thin plume of sand is spread downwind, which, if it falls among plants is not easily re-entrained. Where the ejection of sand is enough, a small free dune may appear, whose size is proportional to the length and depth of the parent blowout, at least on the Manawatu coast in New Zealand (Hesp, 2002b). The bedding pattern in this deposit may be thin sequences of slip-face bedding, disturbed by biotorbation, but more complex patterns are the rule (Neal and Roberts, 2001). A free dune downwind of a blowout may be the start of the formation of a parabolic dune (shortly).

The demography of blowouts is very varied. Most have short lives, which end when plants re-colonise. Between 1973 and 1998 in the Lake Huron dunes, $3991 \, m^2$ of bare sand (mostly in blowouts) was produced, while $4127 \, m^2$ was re-colonised by plants (Dech et al., 2005). A few blowouts grow to a size beyond which they are self-perpetuating (Jungerius and van der Meulen, 1989), which may be the result of funnelling a wide arc of winds through the deepening trough (Hansen et al., 2009). Another threshold size may be passed when widening moderates the constriction of flow, after which the rate of erosion declines, and vegetation may be re-established. A long-term study of two blowouts in Prairie Canada showed that they stabilised when they reached $60 \, m \times 36 \, m \times 8.1 \, m$ (Hugenholtz and Wolfe, 2006). In New Jersey, the cycle from initiation to stabilisation takes ~20 years (Gares and Nordstrom, 1995).

Superimposed on the short-term demography of blowouts, there may be longer-term patterns of growth and decline, as on the Îles de la Madeleine in Québec, where the strongest driver is climatic, especially changes in rainfall ($47°29'00''N$; $61°46'02''W$; $250 \, m$; O'Carrol and Jolicoeur, 2001). In the semi-arid fixed dunes of western Nebraska, blowouts are remobilised in droughts that occur at ~20-year intervals, but probably not without the help of other disturbances, as by cattle ($41°56'35''N$; $101°06'17''W$; $6 \, km$; Stokes and Swinehart, 1997). In northern China, blowouts expand in dry years and contract in wet (Zhang Ping et al., 2008). In some places, relic, re-vegetated blowouts are legacies of windier periods in the past, as in the Duero region of central Spain (Gutiérrez-Elorza et al., 2005).

Parabolic dunes

These were recognised as a category, and named, very early in the study of dunes (Solger, 1908). They reach to between a few hundred metres to over 20 km in length, as on the northern Coral Sea coast of Queensland (the largest fields of parabolic dunes in Queensland are named on Figure 8.1). 'Parabolic' describes the plan shape of the dune, the parabola opening upwind, even if many so-called parabolic dunes have either sharper or blunter apices than a parabola. In many parabolic dunes, there are signs of the migration of a series of dunes down the same path, building up a series of concentric, 'nested parabolic' dunes ($22°52'S$;

150°45′W; 16 km in Queensland, shows a selection of shapes and some nested parabolic dunes). The apices of some parabolic dunes have been destroyed, leaving only linear-dune-like arms (Chapter 4).

Parabolic apices have been found to move with celerities (Chapter 3) of between 0.05 and 30 m yr^{-1}, the fastest rate being usually in drought years, as in the Great Sand dunes in Colorado (Marín *et al.*, 2005). On the Canadian prairies, two monitored parabolic dunes more than doubled in size in 20 drought-dominated years (Hugenholtz and Wolfe, 2006). Clearance of forest after European settlement on Cape Cod in Massachusetts (from about 1690 CE) accelerated the movements of parabolic dunes to 4 m yr^{-1} (measured between 1938 and 1977, a dry period), but they returned to 1 m yr^{-1} between 1987 and 2003, in a wetter period (Forman *et al.*, 2008).

The celerity of a parabolic apex may accelerate if the wind uncovers a hard surface at the centre of the dune, over which any sand entering upwind is transmitted rapidly downwind. Celerity may decelerate where little new sand enters downwind or where the uncovering of a cohesive base or a water table constricts the supply of sand (Hugenholtz *et al.*, 2009). The movement of the apex may also decelerate if vegetation colonises the upwind supply of sand, or the base of the hollow, as could occur in wet or calm periods (Hugenholtz *et al.*, 2008). Many parabolic dunes are thought to have been activated periodically, in dry or windy periods, and re-stabilised by vegetation in intervening wet or calm periods. Reactivation has been revealed by improvements in dating techniques, for example in Idaho (Forman and Pierson, 2003). The vast majority of parabolic dunes are thus relics of the climate of the past (explored more fully in Chapter 10).

It has been suggested for decades that parabolic dunes begin as blowouts. This assertion has some confirmation in Nield and Baas's (2008a) cellular automaton model, which showed just such an origin (Figure 6.5).

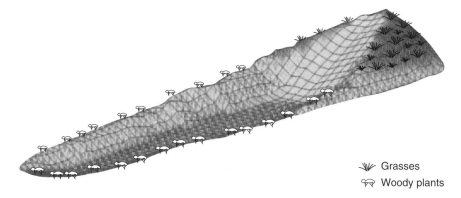

⎯ᾧⳐ Grasses
ᾧᾧ Woody plants

Figure 6.5   Output of a numerical model, which follows the development of a parabolic dune from a blowout; the model includes the effects of the differential survival of two species of plant with different ecological characteristics (Nield and Baas, 2008a), redrawn with new symbols added. Reprinted with permission from John Wiley & Sons.

Nield and Bass's (2010; Figure 6.5) model seems to confirm that a major control on the form of parabolic dunes and even their existence are the ecological requirements of at least two species of plant, a loose-sand tolerant grass on the apex, and plants that need more stability on the arms. Further controls on the development of parabolic dunes are discussed in Chapter 10.

A second model of the development of parabolic dunes supports field observations in which parabolic dunes were seen to originate from the fixture by plants of the low horns of barchans (Chapter 4). In the field, the process seems to take a few decades in coastal dune fields in Denmark and Israel (Anthonsen et al., 1996; Ardon et al., 2009). The mathematical model of this process assumes that: the initial condition is bare sand; plants both increase the roughness of the surface and have various sensitivities to burial; the growth rate of plants is a function of the rate of erosion; in early growth, plants must have access to moisture in the soil in their rooting depth; and lastly, winds are modulated by undulating terrain (changing as a dune emerges). In this model, if vegetation survives at all, it does best on the horns of barchans, where roots have better access to moisture through the thin cover of sand. After the horns have been immobilised, in the model, the barchan or transverse dune is turned inside out to become a parabolic dune. Two interesting predictions of the model are, first, there is a critical, and perhaps sharply defined point, determined mainly by the growth rate of the vegetation, at which a dune turns itself inside out; and, second, the shape of the parabolic dune (sharply pointed or stubby) depends on the balance of the battle between plant growth and sand movement (Durán and Herrmann, 2006). Boat-shaped barchan-parabolic forms discussed in Chapter 10 appear to confirm this scenario.

A third model was developed for the Jockey's Ridge dunes on the outer banks of North Carolina: it showed that a decline in plant cover in interdunes was followed by the development of parabolic dunes (Pelletier et al., 2009b). Thus, field observations and models both suggest that parabolic dunes are initiated in dry spells.

References

Cowling, R.M., Richardson, D.M. and Pierce, S.M. (2004) *Vegetation of Southern Africa*. Cambridge: Cambridge University Press.
Kambatuku, J.R., Cramer, M.D. and Ward, D. (2011) 'Intraspecific competition between shrubs in a semi-arid savanna', *Plant Ecology* 212 (4): 701–713.
Murphy, K.L., Klopatek, J.M. and Klopatek, C.C. (1998) 'The effects of litter quality and climate on decomposition along an elevational gradient', *Ecological Applications* 8 (4): 1061–1071.

Chapter Seven
Coastal Dunes

This chapter, conforming to the scale range of Part II of this book, focuses on an intermediate time-space scale, and on active landforms. The history of coastal dunes is covered in Chapter 11.

Coastal Dunes and Climate

The major distinction of arid coasts, as concerns dunes, is that there is little or no vegetation to hold sand. Where winds have the capacity to move as much sand as is delivered to the beach, which is common, the role of the coast is just another supplier of sand, and dunes close to the shore differ little from those inland (on the north-eastern coast of Venezuela at 12°25′N; 71°36′W; 3 km; and on the west coast of South Africa; 30°32′S; 17°26′E; 28 km; Roberts *et al.*, 2009). But on rare occasions, a pulse of sand delivered to an arid beach exceeds the capacity of the wind to move it further. A coastal fore-dune then develops (unanchored by vegetation), as it did on the hyper-arid coast of northern Chile (Paskoff, 2005a).

A 'humid' climate is one that can support enough plant cover to hold the amounts of sand delivered by the wind from the beach (with a major exception – next paragraph). These coasts share nebkhas (Figures 5.2 and 6.4), blowouts and parabolic dunes within semi-arid environments (Chapter 6).

At some times, even in very humid climates, a large enough pulse of sand is delivered to the beach, and is then blown to the dunes, where it can bury

Dunes: Dynamics, Morphology, History, First Edition. Andrew Warren.
© 2013 John Wiley & Sons, Ltd. Published 2013 by John Wiley & Sons, Ltd.

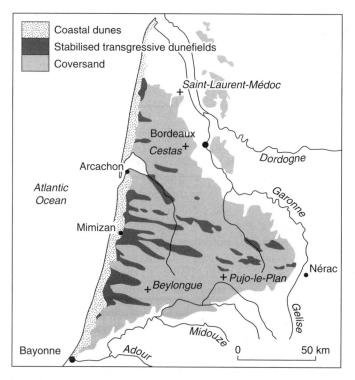

Figure 7.1   Traces of now-stabilised transgressive dunes in the Landes of south-western France. The transgressive dunes were superimposed on an older sand sheet (redrawn and edited; Bertran *et al.*, 2009). Reprinted with permission from John Wiley & Sons.

even vigorous vegetation. These 'transgressive dunes' may reach several kilometres inland (Hesp *et al.*, 1989; Figure 7.1). The pulse may come either from a large quantity of sediment delivered to the beach by along-shore drift or from already-stabilised dunes that have been denuded of vegetation, as is apparently the case on Fraser Island in Queensland, where transgressive dunes, reaching up to 2.5 km from the coast, occur downwind of deflated areas (25°17′55″S; 153°12′57″E; 2 km), despite a mean annual rainfall of ~1800 mm yr^{-1} and a forest canopy that reaches ~40 m high. A family of longer transgressive dunes (now stabilised) is shown in Figure 7.1. Some of these travelled over 50 km from the coast. Given the general directional reliability of most coastal winds (many of them sea breezes; Chapter 4), most dunes on transgressive plumes are transverse.

Some of the first measurements of dune celerity were made on transgressive dunes on the south-eastern Baltic coast (9 m yr^{-1}; Keilhack, 1896); they inspired a classic mathematical model of free dunes (Exner, 1928); and they still attract free-dune research (Badyukova *et al.*, 2007). Another well-studied transgressive dune field is Lençóis Maranhenses in north-eastern Brazil (02°30′S; 42°52′; 20 km; Parteli *et al.*, 2006).

The Beach–Dune System

Coastal fore-dunes, which are the only exclusively coastal dune, and which are described shortly, cannot be understood out of their context in the dynamic, interactive beach–dune system.

The supply of sand to a stretch of coast is determined by the same kind of mixture of processes as determine the supply of sand to desert dunes, but, given the geological youth of most coastal dunes, dependent as they have been for sand released by the Holocene rise in sea level (Chapter 11), they are easier to identify. The first of these is the local coastal setting, for example, its proximity of supplies of sediment, as from a river mouth or the erosion of cliffs (Chapter 13); and the spatial pattern of the coast (and offshore topography), which may funnel or disperse sediment. The second is the tidal range, which is the main control of the width of beach over which sand may be exposed to the wind. The third is the directional pattern of winds (shortly). These processes combine to determine the calibre and rate of delivery of sediment to the beach, and its cross-sectional shape, which in turn determine the rate and temporal/spatial pattern of supply of sand to the dunes. The complex interaction between these processes produces a very diverse, broad-scale differentiation of coastal dunes.

At the scale of an individual beach/dune system, the supply of sand to coastal dunes is more complex. The simplest situation is a gentle slope from shore to dune, on a beach that dries out uniformly, and on which the wind approaches at right angles to the shore. In this case, the width of the beach is reflected in the size of the dunes, which, as it happens, is the pattern revealed by broad-scale surveys, as in a survey, using LIDAR, of English and Welsh beaches and dunes (Saye et al., 2005).

Looking more closely, there is a huge variety of beach–dune configurations. Take first beaches on the shores of macro-tidal seas, most of which are usually wide at low tide. If smooth, these have high potential to deliver sand to dunes, but most do not have smooth cross-sections. A much more common topography is of shore-parallel bars, between which water-filled runnels act as efficient sand traps (Anthony et al., 2009). Wave and tidal conditions determine whether or not the bars merge into the smooth profile that might deliver sand to the dunes. On the wide, macro-tidal sandy beaches of the Pas de Calais in northern France, smooth profiles are rare: they occurred only once during an 18-month survey. These beaches are furrowed on the Google Earth image of March 2012 at 50°57′N; 01°47′E; 5.5 km. The variation in smoothing is spatial as well as temporal: even closely neighbouring beaches on the Pas de Calais differ in their degree and frequency of smoothing and thus in the size of the dunes behind them (Anthony et al., 2006). In northern Ireland, a number of marine processes, acting at the sub-decadal scale, periodically couple or decouple the dune from the beach (Orford et al., 2003). At a much longer temporal scale, the merging of shallow-water bars in periods of rising sea level probably explains the growth in the rate of the associated coastal dunes (Chapter 11).

Furthermore, winds seldom blow in at right angles to the shore. On the Oregon coast, where most beaches run roughly north-to-south, the Resultant Drift Direction (RDD, Chapter 4) is from the north-west or north-north-west. This gives the wind greater fetch across the beach, and thus greater opportunity to entrain and deliver sand, than if it had been directly onshore. On yet more shore-parallel beaches, more sand is blown along the beach than to the dunes (Bauer and Davidson-Arnott, 2003). Dunes on leeward coasts, as at Tentsmuir in Scotland, may lose sand to the beach, which then aggrades (56°23′N; 2°48′W; 3.5 km; Wal and McManus, 1993).

Another complicating set of processes controls the degree and pattern of the wetness of the beach. The position of the boundary between the wet beach which delivers little sand to the wind, and the dry beach which is blowable, is critical and is itself controlled by several factors, such as the slope of the beach, discharge onto the beach of inland aquifers, wind speed, temperature and the length of time over which sand is exposed (determined, among other things by the tidal cycle). It is also influenced by rainfall, which restricts sand movement, often just when winds are strongest (although rainsplash may help to maintain wind-driven sand movement – Chapter 1). Because they are more liable to changes in water levels than oceanic shores, lake shores are more liable to longer-term changes in width, and therefore in their ability to deliver sand to dunes (Davidson-Arnott and Law, 1996).

Yet more complexity is revealed by studies of sequences of events. A lengthy period of offshore winds in Brittany calms the sea and allows a store of sand to accumulate offshore. When such an episode is followed by a westerly storm, sand rapidly reaches the dunes. But when the sequence begins with onshore winds, which allow little offshore accumulation, a westerly storm may attack the dunes (Regnauld and Louboutin, 2002). Some beaches, and at some times, develop 'mega-cusps' (indentations in the shoreline of the order of 200 m long), as (at times) on the Monterey beach in California (36°40′28″N; 121°48′59″W; 500 m). The dunes directly behind the cusps receive less sand and are lower than else-where along the shore. In some cases, the cusps migrate along the shore, dragging with them the pattern of accretion and erosion (Thornton et al., 2007). On the Dutch dune coast, long-wavelength features move down-drift, at anything up to 200 m yr^{-1} (Ruessink and Jeuken, 2002). Thus, each beach–dune system is con-trolled by its own particular combination of processes working together or against each other, at different times and over different stretches of time. A full explana-tion is seldom easy (Orford et al., 1999a).

As in many landforms, it is moderate winds that move most sand from beach to dune (Figure 7.2; Wolman and Miller, 1960, RL). This is because moderate events are more frequent, and more likely to encounter favourable combinations of the other controls. If an infrequent strong wind encounters unfavourable com-binations, it moves little sand. Favourable conditions for the release of sand being as rare as the high winds that might carry it, the two even more rarely coincide. When they do, strong winds create strong waves that attack the dunes, and thus counter any dune-building capacity. For all the rarity of these coincidences, strong storms can buck the magnitude–frequency relationship of sand movement, as shown in Figure 7.2.

Storms like these play a crucial role in determining the existence and shape of coastal dunes. Storm waves breaking at the base of the dunes claw sand back to the beach, and slumps of wetted dune sand may deliver more, as on Irish coastal dunes (Carter, 1990b). Storm attack is chiefly in the winter in temperate climates (Ruz and Meur-Ferec, 2004), and in the hurricane season nearer to the equator (Houser *et al.*, 2008). Erosion during hurricanes on the eastern seaboard of the United States can cause the retreat of the dunes at up to 100 m yr^{-1} (Sexton and Hayes, 1992). More than 240 km of dunes were 'cliffed' by Hurricane Opal in western Florida in, 1995. Some of the sand then taken from the dunes was first transferred to the sea bed at a depth of 12 m, before returning slowly to the beach and so to the dunes (Leadon, 1999). On the Alexandria dune system on the Indian Ocean coast of South Africa, ~12% of the sand delivered to the dunes from the beach each year is returned to the beach by wave attack (Illenberger and Rust, 1988). In many cases, as on the Polish coast, the sand that is taken from fore-dunes by waves is transported down-drift, where it may join another fore-dune (Borówka and Rotnicki, 2001).

Even after extensive damage by storms, the return of sand from beach to dune after a storm may be rapid. On Fire Island in New York State, the rate has reached 1.0 m^3 m^{-1} yr^{-1} and near to 12.0 m^3 m^{-1} yr^{-1} on parts of neighbouring Long Island,

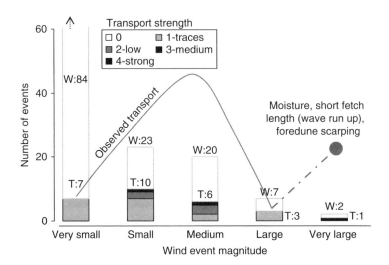

Figure 7.2 Magnitude–frequency effects on sand transport by wind on a beach section on Prince Edward Island in Canada. '*T*' refers to the number of transport events of different strength; '*W*' to the number and strength of wind events. Very large wind events produce high waves that 'scarp' the dunes, but only if they coincide with high enough seas to attack the dunes (redrawn but close to original; after Delgado-Fernandez and Davidson-Arnott, 2011).

which allowed rapid and almost complete rebuilding of the fore-dune (Psuty *et al.*, 2005). But recovery is not assured, for it depends on many factors, such as whether the coast as a whole is prograding or retreating, the strength of long-shore drift, wind directions, offshore slope and so on. Recovery followed a tortu-ous history on a dune system destroyed by hurricanes in New South Wales. A series of low beach berms (or bars) appeared shortly after its destruction, but most were only temporary. When a suitable site was at last established, a new dune developed, persisted for a few seasons and retreated inland towards its former position (McLean and Shen, 2006).

Even when rates of supply are averaged over long periods, all these processes can deliver astonishing quantities of sand to dunes. On strongly prograding (growing, accreting) beaches in the Netherlands, the dunes may receive up to 75 m^3 (m-width)$^{-1}$ yr^{-1} of sand on short stretches, and up to 25 m^3 (m-width)$^{-1}$ yr^{-1} over stretches of the order of a kilometre (Arens and Wiersma, 1994a). Between 1965 and 1987, a 7.8 km^2 quadrangle in dunes on the eastern shore of Lake Michigan (44°04′50″N; 86°28′53″W; 6 km) lost 1,577,592 m^3 but gained 14,570,316 m^3 of sand or, in aggregate, an addition of 12,992,724 m^3 (Brown and Arbogast, 1999).

At the scale of up to a year, and certainly within a season, a beach–dune system is usually highly unstable; at the scale of decades, it may seem stable, or changing at a stable rate; over centuries, most sandy coasts are either advancing or retreat-ing (Carter and Wilson, 1993; Chapter 11).

Exclusively Coastal Dunes

Embryo dunes

Embryo dunes (Figure 7.3) have also been termed 'incipient fore-dunes'. Most embryo dunes are clusters of nebkhas (Chapter 6) gathered round plants that can survive salty conditions and blowing sand (such as *Elymus arenarius* on European dunes). On beaches that yield great amounts of sand to the wind, embryo dunes grow more quickly than fore-dunes (Gomes *et al.*, 1992). Embryo dunes in Japan formed round a creeping grass, with a 28% cover, absorbed 95% of sand being

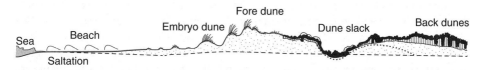

Figure 7.3 Cartoon of a typical cross-section of coastal dunes, with nomenclature used in the text (modified from Arens *et al.*, 2001). Reprinted with permission from John Wiley & Sons.

blown from the beach (Kuriyama *et al.*, 2005). As the word 'embryo' implies, this kind of accretion may lead to growth, even to the creation of a new fore-dune, but embryo dunes are also vulnerable to attack by waves and may disappear and reappear annually.

Fore-dunes ('frontal dunes' or 'retention ridges')

The word 'fore-dune' is occasionally used either for a dune formed upwind of an obstacle or for what is called here an 'embryo dune' (earlier), but more often, as here, it is applied to the main coastal dune ridge.

Where growth is vigorous and the supply of sand sufficient, the fore-dune is a near-continuous ridge parallel to the beach, whose smooth curve it follows. Where plants are less vigorous, as on most arid and semi-arid coasts (or in the past, on coasts to which *Ammophila arenaria* had not been introduced, shortly), the fore-dune is discontinuous and lower. Some of the highest fore-dunes occur on the south-eastern South African coast, where they reach 100 m (Sudan *et al.*, 2004). Many are 10–20 m high. Most are pockmarked by blowouts (Chapter 6). On prograding coasts (earlier), there is a succession of fore-dunes, the older ones further from the beach, progressively more degraded by blowouts or water erosion. Where the coast is rapidly prograding, dunes some distance from the beach may be only a few tens or hundreds of years old. On slowly prograding coasts, they may be up to a few thousand years old, as in the Coorong in Australia (Chapter 11). Maximum rates of upward growth in Irish fore-dunes are between 3 and 4 $m^3 m^{-1} wk^{-1}$ for short periods, and 0.3 and 0.7 $m yr^{-1}$ over longer periods (Carter and Wilson, 1988).

The role of vegetation

On the fore-dune, sand collects round the stems of plants that are adapted to grow upward through accumulating sand. Marram grass (*Ammophila arenaria*) is the commonest of these, having been introduced round the world from its European home in attempts to stabilise coastal dunes. In New Zealand, the introduction was followed by a rapid decline in the area of free dunes (Hilton *et al.*, 2006). There were only fragmentary fore-dunes on the west coast of North America before its introduction (although the degree of activity may have had further explanations). After *A. arenaria* had been introduced, fore-dunes grew and cut off the sand supply to transgressive free dunes (earlier), which then slowly degraded and were colonised by plants (Wiedemann and Pickart, 1996). The North American native, *Ammophila breveligulata*, which had been the main grass on these dunes and dunes on the North American Atlantic coast, is not as vigorous, but has re-invaded some of the western US coast. In other parts of the world, other plants fulfil the dune-building role (Seabloom and Wiedemann, 1994). In humid, warm climates, these include large trees (as on Fraser Island, earlier).

The rate of growth of sand-binding plants depends first on the supply of calcium from shelly material, and nitrogen from other marine debris. Second, it depends on water supply, particularly in semi-arid conditions. Third is the timing of burial in relation to growth periods. Some studies have found that marram thrives in burial rates of $0.25\,m\,yr^{-1}$ and continues growth even at a burial rate of ~$0.6\,m\,yr^{-1}$ (Carter and Wilson, 1990); Dutch studies estimate survival with up to 0.1–$0.2\,m\,yr^{-1}$ (van der Meulen, 1990b). In the right conditions, the roots of newly established marram grow vigorously into new sand, but the roots of marram may suffer 'die-back', which has been attributed to various processes; it seems to be kept in check where there is a supply of new sand (Brinkman et al., 2005 RL).

Aerodynamics

Speedup is less on coastal fore-dunes than on free dunes because of the greater aerodynamic roughness provided by vegetation. This is somewhat compensated for by the acceleration of flow up the steeper windward slopes of fore-dunes, bound, as they are, by plants (Hesp et al., 2009). Speedup on coastal fore-dunes probably has a more complex relationship with the height of the dune than on free dunes. In one field experiment, there was increasing speedup on fore-dunes between 6 and 10 m high, but little more on those between those at 10 and 23 m high, perhaps because of the roughness of the surface. Whatever the speedup, the higher wind velocities at the crest of the dune limit its growth (Arens et al., 1995).

In some conditions, particularly behind steep fore-dunes, flow may separate, as in the lee of desert dunes (Chapter 3). Even without flow reversal, lee-side flow is much more turbulent than in the lee of desert dunes, and large quantities of sand can be carried in suspension and can reach well inland. Lee-side turbulence can take the form of dust devils (whirls of wind that carry sand as well as dust) and downward bursts of wind that carry sands in all directions (Lynch et al., 2010). On lower, less abrupt fore-dunes, even a small lateral component in an onshore wind can create flow in the lee that is accelerated above the velocity of the incoming wind, and be steered parallel to the fore-dune, carrying sand laterally (Lynch et al., 2009). The Ekman dune effect (Chapter 4) may play a role in changing the wind direction between the sea and the dunes.

Internal structure

Cross-bedding (Chapter 3) has been found in sections through a large number of Brazilian fore-dunes and in dunes on Romø in south-western Denmark (Bigarella et al., 2006; Nielsen et al., 2009), but another Brazilian coastal dune, when examined with ground-penetrating radar (GPR), was found to be almost structureless (Buynevich et al., 2010), and a GPR examination of a coastal fore-dune in north Norfolk in England showed that the main features were sets of low-angle beds on the shoreward slope, indicating accretion in troughs such as blowouts (Bristow et al., 2000b).

Mobility/celerity/replacement

Inland migration of fore-dunes occurs on many coasts. Sand is eroded from the seaward face of the dune and taken to the landward face, much as on free dunes (Christiansen and Davidson-Arnott, 2004). One of the main dune ridges at Braunton Burrows in Devon (51°05′45″N; 04°12′42″W; 3 km) moved inland, apparently in this way, by 125 m in 70 years (~1.8 m yr⁻¹) (Kidson *et al.*, 1989). In North Carolina, a coastal dune moved inland at a rate between 3 and 6 m yr⁻¹ (Mitasova *et al.*, 2005). On the west coast of South Africa, a celerity of 5.3 m yr⁻¹ is reported, over a 7.4 ka period ('celerity' as in Chapter 3) (Franceschini and Compton, 2006). Migration makes way for new fore-dunes, as in parts of north-western Ireland, where most fore-dunes do not exceed 3 or 4 m in height before they are replaced by a new ridge (Carter, 1990b).

The exclusion of new supplies of calcium- and nitrogen-rich material in new sand, by the growth of a new fore-dune, causes marram to lose vigour. This allows other plants to colonise what becomes a 'back-dune'. Back-dunes rapidly lose their steep slopes, sometimes because of erosion by runoff (as, sometimes, on water-repellent surfaces; Dekker *et al.*, 2000). The later history of such dunes is covered, briefly, in Chapter 11.

Tsunamis

Tsunamis have left their mark on many dune coasts. Signs of their incursions depend on the size of the tsunami, the time since it struck, and the vigour of local coastal and dune-forming processes. The signs of recent tsunamis include hydro-dynamically shaped remnants of dunes that were breached by the tsunami (as in Otago, New Zealand, 45°43′8″S; 170°36′03″E; 750 m), hummocky topography, being the deposits of the tsunami (Goff *et al.*, 2009), a thin gravelly coating on dunes, a thin sandy deposit on inland surfaces, as in north-eastern New Zealand (Nichol *et al.*, 2004), or landward-thinning sand sheets, as in Oregon and the Andaman coast of Thailand, after the 2004 Indian Ocean tsunami (Kelsey *et al.*, 2005; Choowong *et al.*, 2007).

Coastal sand sheets

Sand sheets extend 6–8 km inland in parts of the southern Cape in South Africa, thinning inland from 6 m deep at the coast (Marker and Holmes, 2002). Most coastal sand sheets are narrower, and are situated directly behind the fore-dune, like the calcareous 'machair' of north-western Ireland and Scotland, especially in the Outer Hebrides (57°15′N; 07°25′W; 3 km; Dawson *et al.*, 2004). Some sand sheets may be degraded remnants of older coastal dunes, as with some inland sand sheets (Chapter 4; Holmes *et al.*, 2007). Some coastal sand sheets may have been formed by breaches of the fore-dune during hurricanes (Nott, 2006) or tsunamis (earlier). The explanation for sand sheets that lie directly on the coast may slow the delivery of sand to the beach (Carter, 1990b).

References

Brinkman, E.P., Troelstra, S.R. and van der Putten, W.H. (2005) 'Soil feedback effects to the foredune grass *Ammophila arenaria* by endoparasitic root-feeding nematodes and whole soil communities', *Soil Biology & Biochemistry* 37 (11): 2077–2087.
Wolman, M.G. and Miller, J.P. (1960) 'Magnitude and frequency of forces in geomorphic processes', *Journal of Geology* 68 (1): 54–74.

Part Three
>0.3 mm; <2,200,000,000 years

Chapter Eight
Sand Seas

Terms

The term 'Sand Sea' (*Sandmeer*) was probably first used by Gerhard Rohlfs in accounts of his gruelling explorations of the Sahara (as in Rohlfs *et al.*, 1875, p. 161). His neglect of a precise definition added to the romance, and turn to the popularity and survival of the metaphor. The alternatives (sands, sand hills, dunes, erg, edeyen, nafu, raml, sha mo or pesky) all suffer the disadvantage of being attached, in local dialects and languages, and on maps by cartographers in their geomorphologically indiscriminate choice of local names, to anything from a single dune to many thousands of square kilometres of sandy country.

Imprecision, happily, does not conflict with the interests of geomorphologists. Size has little influence on the dune patterns in a body of sand, as can be seen in the uniformity of pattern in most of the huge Australian sand seas (Figure 8.1), and the variety in very much smaller sand seas, such as the Coral Pink dune field almost on the Utah–Arizona line (37°03′16″N; 112°42′04″W; 10 km; 7200 km²; Wilkins and Ford, 2007). It is true that, for various reasons, including a limited number of sources of sand, and a more consistent wind regime, smaller collections of dunes are likely to have shorter, less complex histories, but there is no threshold size beyond which history gets complicated.

Numerical definitions withstand analysis no better. They include the inflection in the size distribution of bodies of windblown sand as the lower limit for the size of a sand sea (Livingstone and Warren, 1996, p. 103). The inflexion in that case depended not on any aeolian functionality but on the size distribution of the accommodation spaces of large bodies of sand, most of them tectonic basins. Other

Dunes: Dynamics, Morphology, History, First Edition. Andrew Warren.
© 2013 John Wiley & Sons, Ltd. Published 2013 by John Wiley & Sons, Ltd.

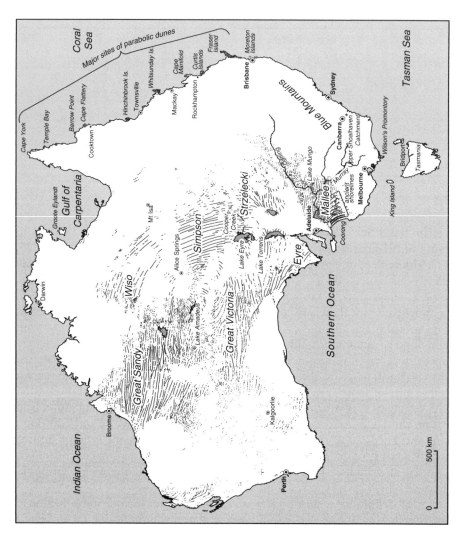

Figure 8.1 Australian sand seas (active, semi-active and stabilised), showing the dominance of sand ridges and the whorl and a half of the driving wind systems (the half whorl is in the West) (thanks to Paul Hesse, 2010).

numerical criteria, such as the proportion covered by sand, the narrowness of the link to other bodies of sand and so on, can only be arbitrary. Attempts at precision would be still more capricious in the case of stabilised sand seas, where the necessary criteria might be required to include such debatable features as the area covered by each phase of growth, and degrees of stabilisation. The rejection of precise criteria in no way invalidates the use of sand seas as a useful term for a body of dunes, or as a convenient unit for the construction of databases, as by Livingstone and colleagues (2010), who also chose not to define sand seas by size.

Large Sand Seas

By any set of criteria, the Rub' al Khali in Arabia (20°N; 49°E) is the largest sand sea on Earth, despite major problems in deciding its exact limits, the extent of its separation from other sand seas and the extent of its inter-dune areas. Other large sand seas include the Great Sand Sea in Egypt (centred at 25°N; 27°E, the objective of one of Rohlf's journeys), the Great Eastern and Great Western Sand Seas in Algeria (30°N; 7°E and 31°N; 1°E), the Kara Kum (40°N; 60°E), the Taklamakan (39°N; 84°E) and the Namib (25°S; 15°E). There are large stabilised sand seas in the Great Sandy Desert in Australia (21°S; 125°E; Figure 8.1), the Sudanese Qoz (14°N; 30°E; Figure 10.10), the Erg Kanem, mostly in Chad and Niger (13°35'N; 12°E; 90 km; located on Figure 10.11), which was the objective of another of Rohlf's journeys, and the Nebraska Sand Hills (42°N; 101°W; Figure 10.6).

Growth and Development

Like all the preceding chapters, this one deals with active processes, albeit ones that take place over much longer periods than development of an individual dune. A measure of the time needed for the accumulation of a sand sea can be taken from a study of the Algodones Dunes in south-eastern California, a small sand sea by global standards, where estimates vary between 750,000 and 160,000 years, depending on assumptions about the rate of supply and its variation over time (Sweet et al., 1988). Such lengths of time inevitably cover changes in climate and other conditions. The yet longer-term development of sand seas is discussed in Chapter 10.

'Active processes' might be stretched to mean the increase in entropy that finally triumphs over short periods of decrease, as dune patterns are consolidated (Chapter 5). Processes that thereafter increase entropy include the slow accretion of area, as in the small Coral Pink dune field referred to earlier; the Kelso Dunes in California, which grew in a similar way (Lancaster and Tchakerian, 2003); and the Wahiba Sands in Oman, which has grown by repeated incursions of marine sand (Warren, 1988b). But neither is there a precise relationship there between time and entropy.

Growth of area usually also involves vertical accretion, which further both increases entropy and builds the mass necessary to survive climatic changes. An example is the Gran Desierto (32°00'N; 113°30'E) in north-western Mexico,

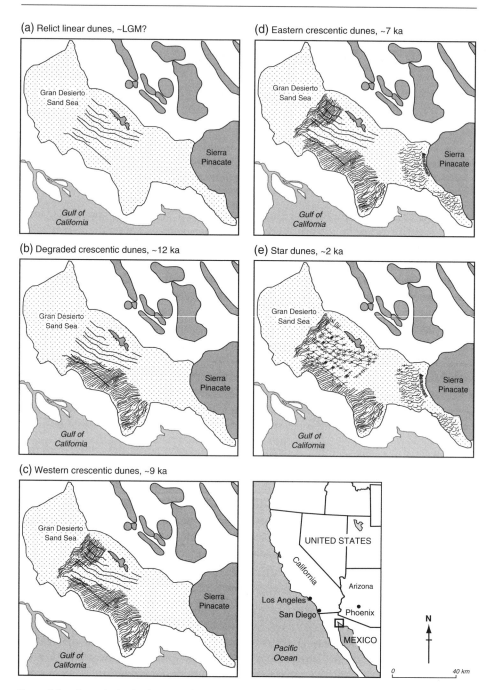

Figure 8.2   Stages in the development of the Gran Desierto Sand Sea in north-western Mexico. Each stage was closed by a 'super surface' (Beveridge *et al.*, 2006). Reprinted with permission from John Wiley & Sons.

which, like many sand seas, holds evidence of a number of dune generations, each with its own dune pattern (Figure 8.2) ('Compound' and 'Complex' dunes; Chapter 10); and the legacy of wet periods, which included the development of soils that either hampered later dune formation or distorted its pattern. Some of the aspects of change at this scale are introduced in Chapter 10.

Sand Seas in Tectonic Basins

Dunes, in general, are features of lowlands: first because basins are aerodynamically smoother and have less violent winds than uplands; and second because most of them hold quantities of the alluvial sand that is the source of most dune sand (Chapter 10). Sand seas in tectonic basins are rendered immobile for long stretches of geological time by the tectonically determined boundaries, even if sand is continually joining from upwind, and leaving downwind. This pattern is best shown by the Idehan Mourzouk (centred on 25°N; 13°E), where sand can be seen entering at 25°38′12″N; 14°32′27″E; 180 km (and other places), and leaving through gaps in the escarpment in the south-west, one at 23°19′N; 12°14′E; 130 km. The association of sand seas and deep tectonic basins is seen elsewhere in the Sahara and in western China, particularly in the Taklamakan (39°N; 84°E), and the Gurbantunggut (45°N; 88°E). If the slopes of the basin are steep, the sand sea is separated from them by a corridor that is swept of sand by turbulence, as in the Issaoane n'Irarraren (26°N; 6°43′E). In that basin, as in many others, the confining slopes are alluvial fans.

Most of the sand seas in other parts of the Sahara, Arabia, northern China, central Australia and southern Africa are also in basins, but shallower ones, allowing for input and output of sand over a wider front. In Australia, there is a debate about the significance of the association. If the reason for the association is the location of sandy fluvial sediments, then there may have been little downwind movement of the sands, perhaps signifying a 'wind-rift' type of dune formation (Chapter 4); if sand moves between basins, then an extensional model of linear-dune formation may hold. A recent survey of the Australian dunes suggests that both models could apply, one here, another there (Hesse, 2010).

In a topographically confined sand sea, the winds that enter the basin lose sand-carrying capacity as they fan out, and thus sand is deposited (Figure 8.3). In the centre of such a sand sea, where winds are least confined, there is less sand-carrying capacity so that dunes build. Where winds are again confined as they leave the basin, usually through narrow passes, they regain speed and sand-carrying capacity. All the elements of this ideal pattern are seldom found in one sand sea. Fanning out is illustrated by the eastern parts of a large sand sea at 27°02′36″N; 14°15′59″E, and reconfinement is seen in one of narrow sand streams leaving Idehan Mourzouk at (coordinates earlier), which itself fans out again in a smaller basin.

Wilson's model also applies to sand seas in which wind speeds falter downwind. Sand then accumulates, and dunes may ride up over each other. This pattern

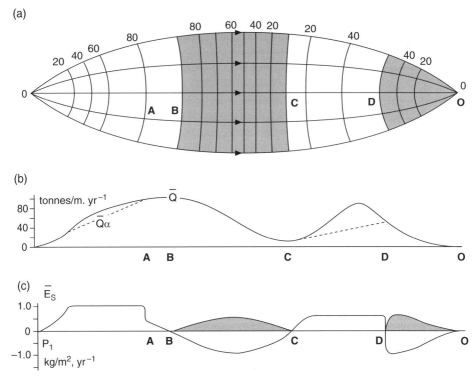

Figure 8.3 Model of sand flow through a sand sea (Wilson, 1971). (a) Map of actual mean sand flow rates $\overline{Q}$ in a stylised sand sea: shaded areas show zones of accumulation, which is partially explained in (b) and (c). (b) Plot of the mean actual mean sand flow $\overline{Q}$ and the mean potential sand transport rate $\overline{Q}\alpha$. (c) Plot of the mean erosion rate $\overline{E}$ (negative $\overline{E}$ = deposition). As Wilson explains more fully, the variations of these quantities depend largely on the convergence or divergence of sand flow, as in (a). Permission from Wiley-Blackwell, Publishers.

applies to the Namib Sand Sea, where sand accumulates at the northern end, after faster travel further south (Lancaster, 1985b). Downwind accumulation has been recorded in many ancient sand seas, now lithified as sandstones (Chapter 10).

Accumulation also occurs when sand is taken to a place where wind speeds, even if high, come from many different directions (in complex wind regimes, Figure 4.3e). Sand then accumulates in star dunes, which may grow to great heights, as in the tectonically confined Kelso Dunes in California (34°55′N; 115°43′W), or in the unconfined Great Western Sand Sea in Algeria (later). Another sand sea that experiences convergent flows of sand is the Gran Desierto in north-western Mexico (coordinates earlier; Figure 8.2; Lancaster, 1990a). Sand is brought in by onshore westerly winds on the coastal areas of the west; by high-energy southerly winds in the south; by topographically steered mountain winds with complex patterns in the east in the centre; and by low-energy directionally variable winds in the north. Sand therefore moves, for example, from the transverse dunes on the coast and in the south towards star dunes in the north and east (Blount and Lancaster, 1990).

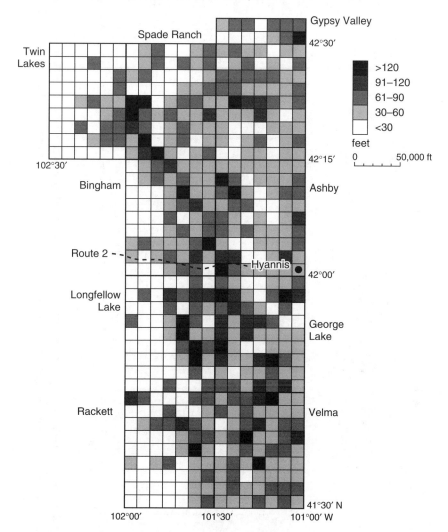

Figure 8.4   Elongated zones of thick sand in the Nebraska Sand Hills, presumably trailing downwind of a major source of sand (after Warren, 1976b).

Topographically Unconfined Sand Seas

Wilson developed his model (Figure 8.3) with unconfined sand seas in mind, not for the confined basins to which it has been applied in the last section (to which it seems to apply pretty well). He mapped winds over the whole

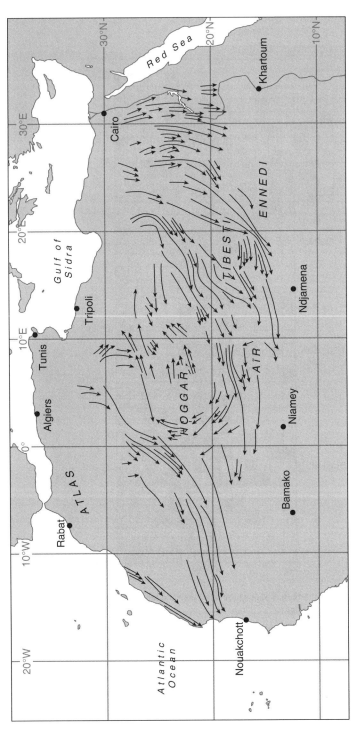

Figure 8.5 Transfer of sand between sand seas in the Sahara (very thoroughly redrawn, edited and put on another projection; Mainguet, 1978). Note the one and a half 'whorls' of the wind system (the half whorl is in the west). Note the apparent transport of sand southward towards the Sahel of West Africa (discussed in Chapter 10).

Sahara, based on the very sparse data available to him. The map showed a compartmental pattern, which included 'peaks', from which sand flow radiated in all directions, expanding, and contracting downwind. It may be that there are unconfined sand seas to which Wilson's model could apply, and the great improvements in both the collection and accessibility meteorological data, and in meteorological modelling of wind patterns, may yet detect such patterns.

But the patterns of Saharan winds that Wilson thought he had found can now be seen (even on Google Earth) to bear little relationship to the dune patterns in most of the Saharan sand seas. For example, Wilson's sand-flow map of the area surrounding and including the Great Western Sand Sea in Algeria showed an arc of flow, which swung from north-eastward in the south to south-eastward in the north, which is wholly at odds with the actual dune trend (in the area surrounding 30°N; 7°E), which follows an arc from the west or north-west in the north, swinging smoothly round to be from the north-east in the south-west (albeit that some of the elements of this pattern are very old, Chapter 10).

Patterns of convergence and divergence of sand flow over the Great Western Sand Sea may well occur, but it conforms better (in broad terms) to Porter's (1986) model of a sand sea, whose development is unconfined by topography, than to Wilson's. Porter's model shows such a sand sea moving as a coherent body, preceded by outlying dunes and trailing some other types of dune. Whether in strict conformance to Porter's model or not, the bodily movement of the Great Western Sand Sea can be confidently inferred from what is now known of its long history (Chapter 10). Other examples of unconfined, or lightly constrained, sand seas include the Great Eastern Sand Sea in Algeria (coordinates earlier); and the Nebraska Sand Hills (41°N; 101°W; Figure 8.4), where the evidence of freedom from topographic constraint lies in the north-west–south-east arrangement of zones of high dunes, which presumably reflect unrestricted transfer downwind from upwind sources, somewhere to the north-west (Figure 8.4).

Transfer between Sand Seas

Many sand seas are linked by regional transfers of sand (as in Australia, earlier). The most extensive transfer is across the Sahara, as interpreted from satellite imagery (Figure 8.5). The pathways of the sand terminate either in the Sahel, where winds are gentler and more multidirectional, and where therefore sand accumulates (Mainguet, 1978b; a pattern whose possible significance is discussed in Chapter 10), or in the Atlantic, where there are deep deposits of windblown sand offshore (Sarnthein and Walger, 1974). Transfer also links some Chinese sand seas, as between the Badain Jaran and the Tengger deserts at 38°N; 104°E.

Patterns in fields of dispersed barchans, which, by the loose definition of sand seas adopted in this chapter, could be called 'sand seas', are discussed in Chapter 4, largely because the distribution of dunes and the shape of the individual dunes in these groups are closely linked.

Chapter Nine
A History of Dune Sand

Provenance

The focus of this chapter is on quartsoze sand, which is derived from siliceous rocks. Carbonate sands, which are less common, are described with the history of coastal dunes in Chapter 12. The yet fewer dunes that have been built of sand composed of gypsum, diatomite, halite, volcanic debris, ice and other substances, some of which form distinctive dunes, are described in Chapter 4.

The conversion of siliceous rock to dune sand is rarely direct. It is at its most direct in Antarctica, where rock is broken down to sand by the crystallisation of salts, but the process is very rare, and the yield, in global terms, is tiny (Bristow *et al.*, 2010a). Very much more sand is taken from rock to dune by very much less direct routes, the most productive of which is by river. Given a sufficient gradient, rivers size-sort the jumble of rock debris that is delivered to them from valley sides, abandon the coarser material upstream, leave sand to the middle and lower reaches, and take the fines further. The grain size of sediments in streams commonly crosses an abrupt threshold from gravel to sand in the downstream journey (Jerolmack and Brzinski, 2010). The change is thought to be largely the consequence of the abrasion of particles in transport (although other processes, like sorting by size, also contribute). The pattern seems to be common, and must be a major control on the location of the transfer of sand from rivers to dunes (Frings, 2007, RL).

A river that carries large quantities of coarse sediment (sand and coarser) flows in many shallow channels, as did the River Tisza in eastern Hungary, during the last glaciation. The Tisza's catchment was then sparsely vegetated and most of the

Dunes: Dynamics, Morphology, History, First Edition. Andrew Warren.

sediment it carried came from coarse glacial and fluvio-glacial debris. Alluvial sand in shallow channels such as these is stored in shoals, which, when exposed at low water, as they are on the banks of the South Platte River in Nebraska (41°04′23″N; 101°59′28″W; 3 km), and as they were on the banks of the Late Pleistocene Tisza, present sand to the wind. By the Holocene, glaciers gone, the Tisza carried less sand and flowed in a single channel, incised into the flood plain, which released much less sand for dunes (Kasse *et al.*, 2010, RL). The history of the Tisza was replicated in many other rivers in north-central Europe, particularly in Poland and northern Germany, and many times throughout the Quaternary (Chapter 10).

In drier parts of the world, the importance of the alluvium of ephemeral rivers and lakes as suppliers of dune sand is seen in the association of long corridors of dispersed dunes downwind of these deposits in the central Mojave Desert of California (Figure 9.1). Alluvial fans, built where rivers broaden out at a steep mountain front, are another source of large quantities of dune sand. The Taklamakan in western China is the best contemporary example of a sand sea most of whose sand came from alluvial fans (37°37′N; 84°00′E; 200 km). Large interior basins, like that in which the Taklamakan is located, have the added

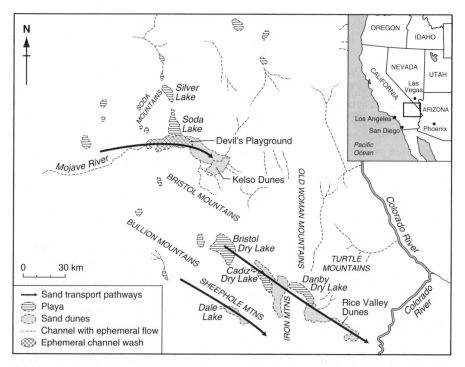

Figure 9.1   Sources of dune sand in the Mohave Desert. Most dunes in the central Mojave derive their sand from alluvium or lake-shore sands (redrawn and edited after Clarke and Rendell, 1998).

advantage, as far as dunes are concerned, that they are very dry, being in the rain shadow of surrounding mountains. The Taklamakan has been claimed as the best modern analogue, size, alluvial fans, rain shadow and all, of Permian Rotliegend Formation, most of whose sandstones are now beneath the North Sea (Glennie, 1983a). The dunes that went to make up some of the ancient sandstones in the western United States also accumulated in deep topographic rain shadows (Chapter 10; Allen *et al.*, 2000). In low-relief deserts, other kinds of multiple-channel river fulfil the same role (25°05'S; 140°21'E; 98 km; Nanson *et al.*, 1995; Figure 9.2) but do not deliver as much sand. The association with rivers is temporal as well as spatial: in the Gila River valley in Arizona, phases of dune formation were closely linked to phases of climatically controlled sedimentation by the river (Wright *et al.*, 2011).

Rivers also play a major role in supplying sand to coastal dunes, over even more protracted routes. A large proportion of the sandy sediment discharged by rivers to the sea is taken by wave-driven drift to beaches, from where it is blown to dunes (Chapter 7). Examples include: the sand from the mouth of the Guadalquivir in south-western Spain, which is first taken to beaches, from where is blown to the dunes of the Coto Doñana (Borja *et al.*, 1999a; 36°54'N; 06°26'W; 23 km); and a somewhat bigger dune field on the Israeli/Palestinian coast that is supplied from beaches fed by sedimentary discharge from the mouth of the Nile (Zviely *et al.*, 2007). There are yet bigger bodies of dune sand downwind of the mouths of the Colorado in the western USA (the Gran Desierto; 32°00'N; 114°23'W; 170 km; Beveridge *et al.*, 2006) and the Orange River (the Namib Sand Sea; 28°41'S; 16°28'E; 480 km; Vermeesch *et al.*, 2010). The relationship between rivers and coastal dunes, like the association with rivers, is temporal as well as spatial: fluctuations in the discharge of sand from the Guadalquivir are followed rapidly by the growth or decay of dunes in the Coto Doñana. Even decadal-scale changes in the discharge of sediment from the mouth of the Tugela River north of Durban in South Africa are reflected (after a lag of a few months) in the growth or decay of down-drift dunes (29°13'S; 31°30'E; 9 km; Olivier and Garland, 2003).

Rivers contribute, but less directly, to two other sources of coastal dune sand. Like the sand taken to the sea by rivers, the sand from the erosion of cliffs cut into alluvium is taken by long-shore drift to beaches and so to coastal dunes (an example from southern California is touched on briefly in Chapter 12). The second and possibly bigger source is offshore, where sand that ends up on dunes comes from the river discharge that evades long-shore drift, or from ancient alluvium inundated by rising sea levels after the end of the Pleistocene, as on the coasts of the North Sea and the Baltic (Klijn, 1990a). The importance of maintaining these supply routes to beaches and dunes in the face of schemes for coastal defence is discussed in Chapter 14.

Lakes supply sand to dunes much as do seas, in some places in comparable amounts (as on the south-eastern shore of Lake Michigan; Chapter 12). Long-shore drift, in lakes as in seas, takes sand from river mouths or cliffs to lake beaches, from where it is blown to dunes, for example at Uvs Nuur in Mongolia

(50°N; 94°E; 200 km; Grunert and Dasch, 2004). In Late Pleistocene North America, the beaches of ice-dammed lakes were the sources of small dune fields, as around the glacial Lake Agassiz (web source: Bluemle, 'Glacial Lake Agassiz', ND Geological Survey). Other lake-shore sands were a source of dune sand in the Mojave Desert (Figure 9.1). Lunettes (Chapter 4) are another, if volumetrically minor, type of dune whose sediment comes from lakes.

Recycling

A high proportion of modern dune sand has been taken from earlier dunes, most of which had already been recycled. Recycling of sand by the wind occurs at many scales. At the scale of a single (large) dune, the sand from older dunes was rebuilt into newer dunes in the Namib (Bristow et al., 2007a). At the scale of a source-bordering dune field (later), Pleistocene dune sand was recycled to feed Holocene dunes in western Texas and eastern New Mexico (Muhs and Holliday, 2001). Many other examples are given in what follows.

Other geomorphological processes also help to recycle dune sand. On the coast, there is continual, high-volume interchange between dunes and waves, at scales from the seasonal to the multimillennial (Chapter 7 and later). Sand is also exchanged between rivers and dunes, again at different scales. At an annual or seasonal tempo, recycling on Banks Island in northern Canada begins when sand, abandoned on the bars and banks of a river, is blown to small dunes. The dunes then migrate downwind and downstream to places where some of them are reclaimed by the river (Good and Bryant, 1985). A cycle of tens of thousands of years takes sand from Late Pleistocene and Holocene dunes to the alluvium of South Platte River and back again (earlier; Muhs et al., 1996). Exchanges with a yet longer rhythm, spanning glacial and interglacial periods, shuttled sand between rivers and dunes in the Later Pleistocene of Poland. Highly 'aeolised' sand (shortly) in Polish dunes of the Vistulian (last) glaciation is believed to have been inherited from glacial outwash streams carrying large quantities of debris, which included sand from earlier dunes (Manikowska, 1991b). The evidence for this recycling lies both in the greater aeolisation and maturity of the dune sands of the last glacial period, than those of the penultimate glacial period; and in the high degree of aeolisation of the fluvial sands dating from the last glaciation (Gozdzik, 2007). Repeated transfer between rivers and dunes has also been discovered in south-western Queensland (Figure 9.2).

Recycling is almost certainly the explanation for the rarity of sands dating from earlier than the Holocene in the Mojave and the Canadian prairies (Muhs and Zarate, 2001; Wolfe et al., 2002a). The result is that, in sequential samples from a pit or a core, there is apparently much more young than older sand (Nanson et al., 1992).

Recycling is probably faster in high latitudes than in low latitudes. In high latitudes, the weakly coherent podzols (spodosols) that develop on the acidic sands of stabilised dunes, as in Upper Michigan, offer little resistance to recycling

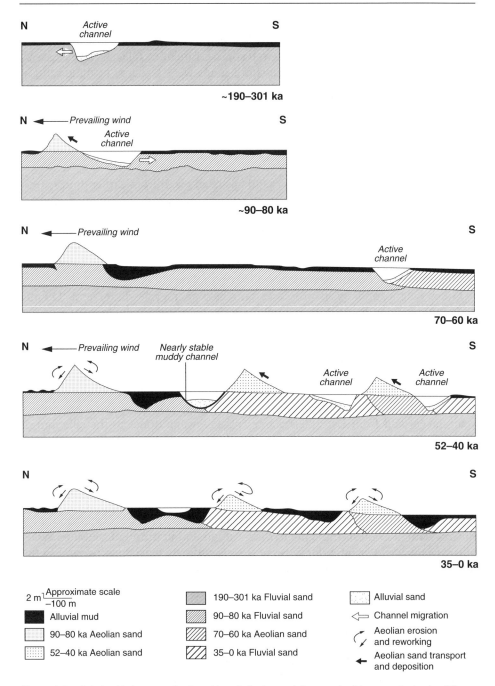

Figure 9.2 Relationship between the deposition of alluvium and the growth of dunes on the banks of Cooper Creek in central Australia, over a sequence of wet and dry phases in the last ~100,000 years (redrawn and thoroughly edited; Maroulis *et al.*, 2007).

(Arbogast and Jameson, 1998). Soil development in the drier semi-arid tropics is rarely continued for long enough to impede recycling, but in the wetter semi-arid tropics, weathering is faster, and quantities of dust may be added to dune surfaces. On the north-western margins of the Australian sandy deserts, the upper horizons of soils developed on stabilised dunes have 20–25% by weight of silt and clay (Goudie *et al.*, 1993). On the south-eastern fringes of the Australian deserts, 0.5–0.7 cm of dust was being added to surface per thousand years in the dustiest periods (Chen Xiangyang *et al.*, 2002). In some places, clay, derived from dust that is added to the surface of soils, is translocated down the soil profile to form tough clayey 'B$_t$' horizons, of varying thickness that put a break on recycling when they are exposed by erosion (Chappell, 1998). Even greater resistance to recycling is offered by calcrete, which has developed in some old dune sands, as in some parts of western India (Dhir *et al.*, 2010).

The Quaternary was merely the latest, and very minor, episode of recycling and accumulation. Recycling between rivers and the wind must also have occurred in the deposition of ancient sediments, as is shown by the close association of aeolian and fluvial sands in the Proterozoic Mancheral quartzite in east-central India (Chakraborty and Chaudhuri, 1993). Over stretches of geological time like these, periods of recycling and accumulation are ended by tectonics. Some contemporary dune sand may even derive, ultimately, from very ancient windblown sandstones.

Maturation

Mineralogy

Mineralogical maturity is a product of recycling but has other implications. Mineralogical maturity is the ratio, in a sand, of mechanically tough, chemically stable minerals (such as quartz and ilmenite) to friable, softer and more weatherable minerals (particularly the feldspars). Besides the maturity they inherit from their parent sediments (by recycling, earlier), maturity in windblown sands is the result of two processes (acting alone or in sequence but not together): abrasion during saltation (producing dust that is blown away); and/or chemical weathering, and subsequent leaching of the more soluble minerals, most of which are also the more friable (Muhs, 2004).

The maturity of a sand is a product of the different weightings of three controls, first the mineralogy of the parent sands; second the length of time and the climatic history between the release of the sand from rock and its lodgement in a dune, and third the climatic history after incorporation into dunes. The nature of the parent sands dominates the maturity trajectory in western Argentina, where most dune sands are volcanic in origin and have spent little time since their release from the parent rocks (Tripaldi *et al.*, 2010). Roughly the same applies to the immature sands of the Taklamakan, although these have inherited a higher proportion of tough minerals than have the Argentinean

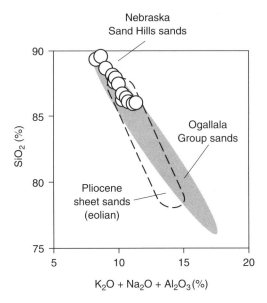

Figure 9.3   Mineral maturity of the sand in the Nebraska Sand Hills compared with the maturity of possible sources. The x axis plots the content of the main chemical constituents of soluble/friable minerals; the y axis plots the silica content (silica being the main constituent of insoluble, hard minerals such as quartz). Mineral maturity is related directly to the content of silica. The possible sources of the sand in The Sand Hills (for example the Miocene to Early Pliocene Ogallala Formation) are less mature than the sand from the Sand Hills. The sands of a later Pliocene sand sheet are a possible intermediary between the two (Muhs, 2004).

sands (Honda and Shimizu, 1998). The Namib sands also reflect the mixed mineralogy of their parent rocks and the alluvium produced from them, because, despite its great age (Chapter 10), the Namib Sand Sea has not suffered periods that were long or intense enough either of moist climates or of the dry, windy climates that might have further matured the sand; thus Namib dune sands are not very mineralogically mature, even if they have spent a long period in place. Inheritance also dominates the maturity trajectory in the Algodones dunes of south-eastern California, but here the sediment is inherited from the alluvium of the Colorado River, whose sandy load was already mature (some of it taken from very ancient windblown sandstones). In contrast to all of these trajectories, heritance has been almost obliterated in the sands of the Nebraska Sand Hills: they are more mature than the sand in any of their possible sources (Figure 9.3). Much of the maturity must have been reached in the dunes themselves, either by abrasion in windblown transport in cool dry periods or by weathering and solution of the dune sands in warmer, wetter interludes or both in some kind of sequence. All this points to a long aeolian history of the sand in the Sand Hills (Muhs, 2004). The maturity trajectory of sand in the Rub' al Khali in Arabia is similar (Garzanti *et al.*, 2003).

Size characteristics

Size characteristics are dubious criteria of maturity, for two very different reasons. The first is the rapidity with which characteristically aeolian size distributions develop in the sediment moved by the wind. One of Bagnold's innovations was a log–log (or log–hyperbolic) plot of size against frequency on which he plotted the change towards a characteristically windblown size distribution, as it blew down his 30 ft (9.144 m) wind tunnel (Bagnold, 1941, pp. 127–143; and in earlier papers). The changes were profound, even in this short journey. In general terms, the wind had selected finer and better-sorted sands. Thus, the plot may be a useful tool for pinpointing the boundary between beach and dune sand in Holocene coastal sediments (Knight *et al.*, 2002a), but rarely discriminates between sediments with longer and more complex histories. The second reason for the dubiousness of size criteria as measures of maturity (or as means of discriminating between windblown and other sands) is much longer term. It is recycling (earlier), which is probably most active on coasts.

Nonetheless, the more salient patterns of size distribution are apparent after some protracted periods of blowing. Thus, the wind has separated fine sands from alluvium in the Taklamakan and carried them further than the coarser fraction (Yang Xiaoping *et al.*, 2007b).

Shape and surface texture

These have also been seen as measures of maturation. The debate about roundness has a history of well over a century. Mackie (1897) maintained that windblown sand was rounder than sand in other sedimentary environments. He boosted his claim with a simple mathematical model of 'rollability' or 'R', where

$$R = dsx / h,$$

in which d is the grain size; s is the specific gravity; x is the distance rolled; and h is the hardness of the mineral.

Mackie derived the distance rolled by some sands by measuring the distance between an outcrop of granite, which he had good reason to suppose was the origin of the sand in what became an aeolian sandstone in Morayshire (an outcrop of the Devonian Old Red Sandstone). He found that $R = 0.28$ for quartz sand and 1.2 for muscovite.

But Mackie did not measure roundness, and a scale of roundness did not appear until over 50 years after the publication of his paper. This was Powers' (1953) scale, which can be used to measure two criteria of shape, namely 'roundness' (of edges and protuberances), which is a preliminary to the rounding of the whole grain, which is its 'sphericity' (Barrett, 1980). The analysis of roundness using Powers' scale is visual: photomicrographs of individual grains are compared with a series of graded templates. Fourier analysis of roundness, also measured from blown-up photographs of individual grains, though even more tedious, is probably a better method of analysing shape (Mazzullo *et al.*, 1986).

Neither did Mackie consider the possibility that there might be selection for shapes in wind transport, and later research has not produced a clear answer. One study found that spheres were lifted more readily than less spherical particles, perhaps because angular grains are held more securely in the bed (under water; Komar, 1987, RL). Another found that angular sands were moved preferentially, perhaps because they protruded further above the bed (Stapor *et al.*, 1983). As shapes become more complex, so does their behaviour in the wind (Chapter 1).

Roundness and surface texture can be lumped together as 'aeolisation', with the justification that they are both outcomes of abrasion in transport, which both rounds sand and covers its surface with 'upturned plates' (or 'frosts' them) (Figure 9.4). Aeolisation has proceeded, sporadically, for hundreds of thousands of years in some areas, as throughout the Quaternary in Poland, where it can be used as a tool in stratigraphy (Mycielska-Dowgiałło and Woronko, 2004).

There is now little doubt about the diagnostic value of upturned plates as evidence of transport in the wind. They have survived legal scrutiny in forensic investigations into the provenance of grains (Culver *et al.*, 1983), even if there are some provisos about their diagnostic value in older sediments (shortly).

The debate over the last few decades seems to have reached two conclusions. First, roundness is not a good criterion for differentiating windblown from beach sands on coasts (Mazzullo *et al.*, 1986). The first reason is probably that the distance from beach to dune is too short. The importance of distance was confirmed by studies in which and grains were blown round in a circular wind tunnel (Kuenen, 1960a). The second reason is probably the mixed environmental inheritance of both beach and coastal dunes sands, because of their constant recycling, at many scales (as was noted earlier in respect to size–frequency differentiation).

Another conclusion about shape as a criterion of windblown sands is that coarse and high-density grains are the most rounded. Kuenen found no rounding at all in grains <0.5 mm after they had been whirled round his wind oval tunnel, and this corroborates Mackie's and Dott's (2003) observations on ancient sandstones. Dott claimed that the particle-size range chosen by Goudie and Watson (1981) for their measurements of roundness, where they found little rounding in a widely drawn sample of windblown sands, was too fine.

There are still mysteries about roundness. Why, as Mackie found, do hard and soft minerals become more rounded above the same grain size? Second, why in a spectrum of grain sizes, is there a sharply defined critical size (or sizes) above which grains suddenly become rounder (Tanner, 1956)? Is this related to or comparable with the sudden downstream jump from coarse to sandy load in river sediments (earlier)? And why do very ancient aeolian sandstones contain the roundest of sands? Is this because of the absence at that time of terrestrial vegetation (Dott, 2003)?

Figure 9.4   Scanning electron microscope images of 'aeolised' sand grains. (a) A grain rounded by collisions in windblown transport. (b) Conchoidal fracturing (produced experimentally at high energy) creating the upturned plates that are characteristic of windblown activity. With kind permission from Peter Bull.

All these findings and debates underline the hazards of relying on a few diagnostics, like size distribution or shape, to discriminate between windblown and other sands, especially in ancient sandstones, but even in Late Pleistocene sands. If well developed, good sorting and roundness may be indications of an episode of movement by the wind, but they could have been acquired either before or during their final deposition. In Poland, the abundance of rounded and 'frosted' sands in some Pleistocene fluvial sands were inherited from earlier dune sands (Mycielska-Dowgiallo and Woronko, 2004). A much older formation, the Ordovician St. Peter Sandstone, which outcrops here and there between Missouri and South Dakota, was thought at one time, and by some authorities, to be wind-blown, largely because of the size and shape characteristics of the grains, but is now thought to have a fluvial origin (Dott, 2003). None of these criteria, there-fore, is a reliable measure of maturity in a windblown environment, unless cor-roborated by the other criteria discussed in this section.

Redness

The colour of dune sands can reach a Munsell colour of 10R, and the degree of redness is yet another candidate as an indicator of the maturity in dune sands. Redness is directly related to the content and degree of crystallisation of haema-tite in the coating of grains, haematite being a product of the weathering of a number of iron-rich rock minerals, most commonly magnetite (Bullard and White, 2002).

Redness has at least six controls. The first and usually dominant control is provenance. The redder sands in the northern Namib Sand Sea and the north-eastern Rub' al Khali (23°N; 54°E; 350 km) come from terrestrial sources, while the greyer sands in the north-eastern Rub' al Khali come from marine sources (Walden et al., 2000; White et al., 2001). The supply of iron in dust is a second control, which may occasionally take precedence, as it may have for the remark-ably red dune sands of northern Arabia. Three bits of evidence suggest that the redness of these sands came from dust: the sands beneath the red coating had already been rounded; the redness resides wholly in a coating on the grains; and the coating contains as much iron as does the grain beneath (Phillips, 1882). A third factor is the time since deposition, as in the Hamra (red in Arabic) soils of the Israeli coast, which become redder with distance from the sea, which is a measure of their age since they were on the beach (Levin et al. 2007a). The dated Mauritanian sequence described in Chapter 10 has also reddened with age.

Downwind reddening (reported by Norris, 1969, among others) could be age related if the sands came from a reducing environment in wet alluvium, and were therefore grey, and had then travelled downwind in an oxygenating windblown environment, reddening the while (Wasson, 1983a). Downwind reddening might also occur if abrasion in windblown transport exposed magnetite to oxidation. A fourth process may have reddened windblown sands on Mars, where the atmos-phere contains much less oxygen: the mechanical tumbling of sand like the Martian sand, which subjects it to repeated abrasion, for 212 days (on Earth) produces haematite, apparently independently of chemical oxidation (Merrison et al., 2010). On Mars, and perhaps on Earth, a fifth possible process possibly related to the electrical discharges in dust devils could also have caused oxidation (Delory et al., 2006), although this must be a very minor part of the explanation for redness on Earth. The sixth factor is climate: a warm and moist climate favours reddening (Walker, 1979). The climate need not be the contemporary version: redness could have been inherited from a wetter, warmer period, as Folk (1976a) proposed for the extraordinarily red colour in many central Australian sands (24°S; 137°E; 150 km).

In sum, redness may be related to maturity, but the relationship is nowhere simple, and there can be no universal redness–age relationship. Only if some of controls are held constant (provenance, distance from source, input of dust, climatic history), as they might be in a restricted area, can age be inferred from colour.

Relationships between Dune Fields and the Sources of Their Sand

The following scheme builds on one developed for north-central China (Han Guang *et al.*, 2007).

Source-bordering dune fields

Source-bordering dune fields (which may be followed downwind by 'dune plumes') are built downwind of their primary sources sand (Han Guang and colleagues subdivide this category). The narrowest plumes take sand from the inner curves of single meander bends, as on the western banks of the Tigris in Iraq (Al-Janabi *et al.*, 1988), or from the beaches of small coves, as at 30°35′S; 17°26′E; 12 km on the west coast of South Africa. Some plumes reach great lengths. In the Mojave Desert in California, a plume begins at Emerson Playa, climbs ridges up to 100 m above the plain, falls down their leeward slopes and ends at the Colorado River 130 km from its origin (Figure 9.1; Zimbelman *et al.*, 1995). The plume trailing eastward from Uvs Nuur Lake in Mongolia (earlier) reaches 150 km from the lake and, like the Californian plume, garners sand from subsidiary sources along the way. Bigger source-bordering dune fields derive from sources that are wider across the wind, such as a succession of meanders, the successive positions of migrating meanders, or wide lake or sea beaches. The sand in this category of dune fields bears a close mineralogical resemblance to sand in the parent deposits, as in Iraq (earlier).

Dune fields that have migrated away from their source

The source of this type of dune field is either a river or a beach, but migration has occurred as the supply of sand has declined. Because such a decline is usually associated with a climatic change, most of these dune fields are ancient. The Algodones dune field in California has moved at 16.6 m per 1000 years away from its source in beaches of Lake Cahuilla, a precursor to the Salton Sea (33°N; 115°06′W; 86 km; Stokes and Swinehart, 1997). There are further examples of this kind of migration in windblown sandstones (Kocurek, 1996).

Sand seas that have taken sand from many local sources

These accumulations are yet older, their sands yet more mature (earlier) and less easy to relate to their fluvial sources. During the 40-million-year history of the Taklamakan sands, the main size fraction of dune sands is a mixture of sand from widely separate alluvial sources (Yang XiaoPing *et al.*, 2007b).

The Australian sand seas and some aeolian sandstones

These sand seas do not fit as neatly into this classification. Their sand came ultimately from igneous rocks that are now very distant, and was mobilised and deposited as alluvium many times before it was built into dunes. The Australian dune sands are one such example. Analysis of their uranium/lead (U/Pb) ages shows they are mixtures of material from several 'proto-sources', some far apart, and many far from the dunes themselves, among which are sources in Antarctica, which have been separated from Australia by plate tectonics. Some of the fluvial routes from proto-source to dune in Australia have been comprehensively disrupted by geomorphological and geological changes over a great span of geological time (Pell *et al.*, 2000). The Permian/Jurassic sandstones of the Colorado Plateau in the western USA are another example of this category. Their progenitor igneous rocks are also far removed from the sandstone itself, because, like the sources of some of the Australian sands, they have been moved great distances by plate tectonics (Dickinson *et al.*, 2010b).

References

Frings R. (2007) 'From gravel to sand: downstream fining of bed sediments in the lower river Rhine', *Nederlandse Geografische Studies* 368: 23–56.
Kasse, C., Bohncke, S.J.P., Vandenberghe, J. and Gábris, G. (2010) 'Fluvial style changes during the last glacial–interglacial transition in the middle Tisza valley (Hungary)', *Proceedings of the Geologists' Association* 121 (2): 180–194.
Komar, P.D. (1987) 'Selective grain entrainment by a current from a bed of mixed sizes: a reanalysis', *Journal of Sedimentary Petrology* 57 (2): 203–211.
Zviely, D., Kit, E. and Klein, M. (2007) 'Longshore sand transport estimates along the Mediterranean coast of Israel in the Holocene', *Marine Geology* 238 (1–4): 61–73.

Chapter Ten
A History of Inland Dunes

This chapter begins with a short account of a very long span of dune history (before the Quaternary). The core of the chapter is a discussion about dune historiography, which explores the degree of confidence that can be placed in what follows, which is a longer account of the dunes of a much shorter period (the Quaternary).

Very Ancient Dunes: Siliceous Windblown Sandstones

Sandstones are compacted and cemented (lithified) sands (Figure 10.1). Windblown sandstones are lithified dunes, known also as 'aeolian arenites' and 'aeolianites'. Only quartz-rich sandstones are described here. Calcareous sandstones, being almost exclusively coastal, are covered in Chapter 11.

The first quartz-rich windblown sandstone, and hence the primordial dune, may have formed 2,200 million years ago. The absence of earlier dunes from the record could be: because stream channels were unconstrained by vegetation before that date, and thus unable to deliver sand in the quantities needed to form dunes; or because there was repeated marine reworking; or, most humdrum, because none has yet been found (Eriksson and Simpson, 1998).

Most bodies of aeolian sandstone, like most contemporary dune fields, are small. Their parent dune fields drew on limited sources of sand or occupied cramped accommodation spaces. An example is the Permian sandstone in north-western England and south-western Scotland, which now buries, is buried by, is sandwiched between or is juxtaposed with fluvial sandstones. Tectonic movements have further dismembered this and other aeolian sandstones (Brookfield, 2008a).

Dunes: Dynamics, Morphology, History, First Edition. Andrew Warren.
© 2013 John Wiley & Sons, Ltd. Published 2013 by John Wiley & Sons, Ltd.

Figure 10.1   Outcrop of the Early Permian Coconino Sandstone on the South Rim of the Grand Canyon.

Much fewer of the surviving bodies of ancient sandstone are large, also like dunes today. These are the vestiges of sand seas in the deserts of ancient mega-continents. The oldest of these accumulated on Rodinia, the Neoproterozoic mega-continent. Their remnants were later scattered by plate tectonics, one to the San Bernardino Mountains near Los Angeles; others to South Africa and India (Biswas, 2005; Stewart, 2005). One of the sand seas of Pangea, the next mega-continent (in the Palaeozoic and Mesozoic), was the 'Chuska Erg', which accumulated for half a million years. Like the others, its lithified remains were then dispersed, one to straddle the northern end of the Arizona/New Mexico border, where this fragment alone covers 140,000 km^2 (more than England), and is up to 535 m thick (36°23′N; 109°00′W; 3 km, view in 3D; Cather et al., 2008). The dune legacy of the Gondwana mega-continent (the most recent) was also torn apart, taking the Botucatú sandstone to Brazil and the Etjo Sandstone to Namibia (Mountney and Howell, 2000; Scherer and Goldberg, 2007; Figure 10.2).

A subsiding basin provided the best conditions for the accumulation of a body of windblown sand. In such a basin, sand built up as dunes rode over each other

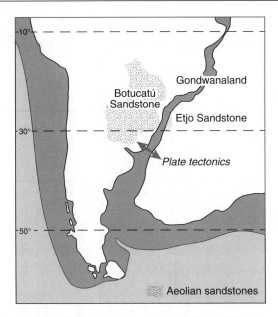

Figure 10.2   Map showing Etjo and Botucatú Sandstones. Etjo and Botucatú Sandstones derive from the same sand sea on the mega-continent of Gondwana but have since been separated by plate tectonics, one going to Namibia, the other to Brazil (edited and redrawn, after Scherer and Goldberg, 2007).

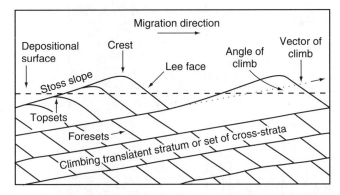

Figure 10.3   Climbing structures and slip-face bedding (or cross-strata) in aeolian sandstones (Kocurek, 1991). Reprinted with permission from John Wiley & Sons.

along 'accumulation surfaces' (Figure 10.3). These, and other breaks in growth, created a hierarchy of surfaces, the most distinct of which are multimillennial erosional interruptions or 'super-surfaces' (as in Figure 10.4). In some bodies of windblown sand, super-surfaces may have been created to the rhythm of the orbitally forced changes in solar input of their time (of which more later)

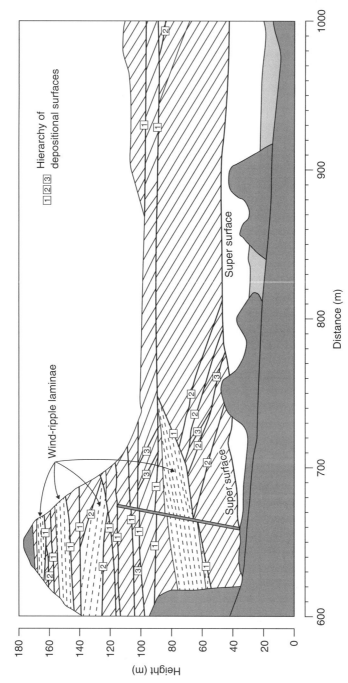

Figure 10.4 Cross-section through the Etjo Sandstone in Namibia, showing a hierarchy of surfaces numbered in ascending order of their frequency. The least frequent is the 'super surface' that underlies the whole body of this sandstone. Some beds are composed mostly of pin-stripe laminae, whose significance is explained in the text (after Mountney and Howell, 2000). Reprinted with permission from John Wiley & Sons.

(Clemmensen *et al.*, 1994). Some ancient dunes survived long enough to experience changes in wind direction, either as the climate changed or as they were swivelled by plate tectonics, as was the Early Triassic Bohdasin sandstone in the Czech Republic (Ulicny, 2004). Geologically recent examples of accumulation are the dunes in the Gran Desierto sand sea in north-western Mexico, in which three super-surfaces separate one Late Pleistocene and three Holocene depositional episodes, each lasting ~25,000 years. The deposition of the last episode began ~3000 years ago.

The best diagnostic of windblown origin is 'pin-stripe' lamination, created as ripples pass over each other and build a stratum of pin-stripes (Chapter 2), or by the accumulation on the slip faces of dunes (Chapter 3). Pin-stripe beds in the Etjo Sandstone (Figure 10.4) are the least ambiguous evidence of its windblown origin. The limitations of the sedimentary criteria that have been used to distinguish windblown sand are discussed in Chapter 9.

Authenticated windblown sandstones have most of the features of active dunes: hierarchies of dune size (Chapter 3), as in the Tensleep sandstones of Wyoming; transverse dunes, as in the Tsonab sandstone in Namibia; zibars, as in the Egalapenta sandstones in India; and sand sheets as in the Cretaceous Merilia formation in Brazil (all these dune prototypes are described in Chapter 4; Kocurek *et al.*, 2000; Biswas, 2005; Basilici *et al.*, 2009). The claim that there were few remnants of linear dunes in windblown sandstones (Rubin and Hunter, 1985) was probably premature. Rubin and Hunter noted that few of the bedding patterns associated with linear dunes had been found in aeolian sandstones and proposed that a discrepancy would disappear if 'linear dunes', ancient or modern, had moved sideways, thus appearing as transverse dunes in the rock record. Although sideways movement of linear dunes has now been observed in some active linear dunes (Chapter 4), there are now more records of the bedding patterns of linear dunes in windblown sandstones, as in the Botucatú sandstone in Brazil (Figure 10.2), and in the Bhander sandstone in India (Bose *et al.*, 1999; Scherer and Goldberg, 2007). The shortage of linear dunes in the rock record, if it exists, could also be because linear dunes do not accumulate as readily as transverse dunes.

The Emergence of Familiar Spatial and Dynamic Patterns

By the Pliocene, between 12 and 2 million years ago, the main bodies of windblown dunes had been moved close to their present global positions. The continents have moved less than 100 km since then, some much less. Thus, most of what is now desert (where dunes are common) was already desert in the Pliocene; a far smaller area, now desert, had yet to become desert; and other small areas, then desert, are no longer desert.

As the environments with potential for dunes slid into place, so did the large-scale rhythms of their behaviour. Much the strongest and most prolonged of these are orbitally forced. For many millions of years before the Pliocene, these rhythms had ensured that much of Earth experienced long periods of warm, humid climate. In the Late Pliocene and Quaternary, they combined to intensify and increase the frequency of cold/dry periods, with a period of ~100,000 years, and these became the strongest rhythms of change in windblown environments.

Shorter climatic rhythms in dune behaviour also appeared in the Pliocene. Although not orbitally forced, they have been amplified or dampened by the orbitally driven change (Wanner *et al.*, 2008, RL). Two of these were inaugurated by the closure of the Panama Gap between North and South America about three million years ago, in the Upper Pliocene. The first, in the Pacific, is the El Niño–Southern Oscillation (ENSO), which intensified between 1.9 and 1.7 million years ago and now oscillates with a period of about five years. Many contemporary windblown processes pulse to the ENSO's beat, such as: the periodic enhancement of dune-building winds in central Australia; arid dune-building conditions in semi-arid and arid North America (later); and on the Oregon coast (Hesse, 2010; Ruggiero *et al.*, 2005). The North Atlantic Oscillation (NAO) is a weaker and less far-reaching post-Panama rhythm. Its period of oscillation is irregular, at somewhat shorter than a decade. The NAO has driven cycles of coastal dune formation on the Atlantic shores of Europe, and further afield, ever since (Chapter 11). Other oscillations, as in the Antarctic and Indian Oceans, also date from this geological period.

As far as dunes are concerned, abundant sand was another indispensable legacy of the Tertiary. The warm, humid Oligocene left deep mantles of disintegrated rock. In the cooler, drier climate later in the Tertiary, these mantles lost their protection of vegetation and, where there was accentuated enough relief, were eroded. As aridity further intensified, the sandy alluvium that this created was blown to dunes (Swezey, 2009).

In Libya, large quantities of dune sand came from the alluvium of Miocene rivers that drained northward from the central Saharan highlands. In the Late Tertiary–Early Quaternary, the courses of these rivers were blocked by lavas, and prospective dune sand collected in the sedimentary basin that was formed upstream of the blockage (Drake *et al.*, 2008, RL). In Egypt, Pliocene rivers took debris from the highlands on the rising western flank of the Red Sea rift, westward over the present Nile, whose course had yet to be established, to western Egypt, where it was joined by rivers from the south-west (revealed by radar imagery) (El-Baz *et al.*, 2000). The alluvium of these river systems became the source of the sand in the Great Sand Sea (McCauley *et al.*, 1997, RL; Embabi, 1998). In south-central Arabia, the main Tertiary river system was the Wadi ad Dawasir, which flowed from the central highlands towards the eastern end of the Arab/Persian Gulf (Anton, 1984). A snippet of the middle reaches of the wadi is now revealed by a line of centre-pivot irrigation systems at 24°00′N; 49°07′E; 140 km. Most of the middle and lower courses of the Dawasir and its tributaries are now buried by the dunes of the Rub' al Khali, built with the sand fraction of the Dawasir's own alluvium (Dabbagh *et al.*, 1998, RL; Garzanti *et al.*, 2003).

Further east, the deposits of smaller rivers, flowing south from the Oman Mountains, contributed sand to the Wahiba Sands (22°03′N; 58°00′E; 150 km; Maizels, 1987). The interiors of southern Africa and Australia also began to dry out in the Mid-Tertiary, and dune sands began to accumulate there in the Late Pliocene (Haddon and McCarthy, 2005; Hesse, 2010).

Two large, contemporary sand seas had already begun to accumulate in the Tertiary, both in areas in which aridity also developed early. In the Namib, dune sand has recently been dated with cosmogenic nuclides to one million years ago (Vermeesch et al., 2010), but stratigraphy suggests that there were dunes in parts of the Namib as early as the Miocene or even the Cretaceous (Ward et al., 1983; Segalen et al., 2004). In the Taklamakan, where aeolian activity intensified in the deepening and lengthening rain shadow of the rising Tibetan Plateau, magnetic signatures and stratigraphy suggest that dunes first appeared there at least seven million years ago, also in the Miocene (Sun Jimin et al., 2009).

Dune Historiography

The dune archive for the Quaternary differs in three critical respects from the archives of earlier windblown sands. It is datable by many more techniques, with much greater accuracy, and its evidence is much more geomorphological than the earlier archives, which are largely sedimentary and stratigraphic. What can this archive deliver?

Dating

The biggest weakness of dune archives, and not exclusively those of recent events, is their mobility. But as concerns their use over periods of more than two or three millennia, transverse dunes erase the record at rates of up to once in two decades (data on movement are given in Chapter 3); linear dunes erode and deposit on alternate flanks, usually on an annual rhythm. Sand sheets (Chapter 4) appear to be insensitive to climatic wetting or drying, remaining active, and thus destructive of their archive, for long periods. The best archives in active sand are held in the large forced dunes that build only upward, but they are scarce (Chapter 5; Chase, 2009). Mobility is not confined to active dunes. On the margins of the desert, the activity of dunes is very sensitive to small fluctuations in rainfall and wind speed, well documented in the south-western Kalahari (Stone and Thomas, 2008). The best archives are held in well-stabilised dunes in areas that are now in humid climates, the date they yield being the time since immobilisation (although, even here, there may have been episodes of remobilisation). But, even in long-stabilised sands, bioturbation (mixing by burrowing animals and disturbance by roots) may have jumbled some of the record (McFarlane et al., 2005).

Morphological dating

Before the discovery that carbonaceous material could be dated by the ¹⁴C method, nearly 70 years ago, morphological methods were all that was available for the dating of dunes. But morphological dating has now been revived.

Some methods of morphological dating capitalise on the very mobility of free dunes. One of these was applied to the narrow stream of barchans in the Abu Moharik dune field in Egypt (between 28°00′N; 29°32′E and 26°15′N; 30°13′E; both 8 km). Dividing the length of the dune field by the rate of advance of its constituent dunes gives the whole dune field an age of 35,000 years ago (Ball, 1927a). Bagnold (1941, p. 219) applied Ball's method to another stream of barchans further south, giving him an age of 7000 years, which, he noted, coincided, approximately, with a contemporary estimate of the date of the presumed increase in aridity and consequent abandonment of the Western Desert by pre-dynastic folk. The last major desiccation of the Western Desert is now thought to have begun ~5000 years ago (Haynes, 1982), perhaps within Bagnold's error bars. Much more recently, the same kind of calculation was made for another narrow dune field on the south-eastern side of the Wahiba Sands in Oman. It suggested that the group of dunes set off from the coast, 48 km away, ~4300 years ago. Their sand could have been liberated from cliffs by the rising seas after the end of the last glaciation (which ended ~8000 years ago), which could then have been taken by long-shore drift to the beaches upwind of the present position of the dunes (Warren, 1988b; perhaps as at Liwa in the UAE, later). Applying another method that uses the mobility of dunes, this time to a single dune, Tsoar estimated the age of his seif in Sinai, by dividing the rate of its downwind extension into its present length. This gave it an age of ~250 years, which coincided with the start of a period of intense grazing by domestic stock (Tsoar, 1995b).

Another method of morphological dating gives only relative ages. This is the recognition of 'Compound' and 'Complex' dune patterns. A compound dune pattern is one in which there has been the superimposition of two dune systems of the same prototype (as of a linear on a linear or a transverse on a transverse dune). The superimposition is revealed by a difference in size or orientation between the superimposed and superimposing dunes, as in El Djouf in eastern Mauritania at 20°32′N 07°46′W; 19 km, where the seifs are clearly both younger than the large linear dunes, and created by a different wind-directional regime. A complex dune pattern is one where there is superimposition of dissimilar prototypes (as of linear over transverse dunes, as at 27°08′N; 13°29′E; 45 km in Libya) (Breed and Grow, 1979). The most obvious inference from superimposition is that it was caused by a change in the direction (and perhaps the strength) of winds. This is best illustrated by the compound pattern of the dunes in the Azefal/Agneitir/Akchar dune fields in Mauritania (later) and by the superimposition of seifs on large linear dunes in the Great Sand Sea in western Egypt (also later). In some cases, inferred superimposition is unarguable, even without corroboration by numerical dates (as in southern Libya at 24°29'N; 19°58'E; 100 km, or 18°52'N; 12°43°E; 50 km, and many more examples).

But the assumption that any superimposition is evidence of a change in the wind climate is challenged by the finding of numerical models of pattern formation (Chapter 4). Observations of superimposed small dunes on barchans (Elbelrhiti *et al.*, 2005), the cellular automaton of Eastwood and colleagues (2011) and the numerical model of Schwämmle and Herrmann (2004) all show that small dunes of the same prototype can be superimposed on larger ones, without a secular change in wind directionality or strength.

Pattern analysis

Pattern analysis (Chapter 4) has revived morphological dating. Four examples show its potential. The first concerns some Australian sand ridges that are aligned at 002° (N2°E), whereas the modern wind resultant (the RDD on Figure 4.3) is 158, a difference of 24°. Using data on the size of the dunes and their likely rate of movement, it has been calculated that a change in wind direction 4000 years ago could have caused the realignment, a date that agrees with what is known about the history of the dunes (Werner and Kocurek, 1997). Second, in the Gran Desierto in north-western Mexico, there is a consistent relationship between the parameters of pattern and their known or inferred ages (Beveridge *et al.*, 2006; Figure 8.2). Third, crest-length and spacing data (defined on Figure 4.7) for White Sands in New Mexico, the Algodones in south-eastern California, and a part of the Namib sand sea, match independent estimates of age, the youngest being White Sands (Ewing and Kocurek, 2010b). The fourth case, the Azefal/Agneitir/Akchar dune fields in Mauritania, is examined later, in its regional context.

Numerical dating

The discovery of the ^{14}C technique was a huge leap towards more accurate dating of the recent geological past, but, although well suited to the dating of organic-matter soils within now-stabilised inland and coastal dunes, the severe shortage of carbon limits its application in the deserts and their margins. Even where there is datable carbon, the method loses reliability if the deposit is older than ~60,000 years. The amino-acid racemisation dating technique also relies on organic matter, and like ^{14}C, it has given good results for coastal dunes or buried soils and has a much longer reach (a few hundred thousand years, depending on ambient temperature; Murray-Wallace *et al.*, 2010). But both methods are compromised by the possibility that organic matter has been inherited from sources predating the site in question, or by its mobility within the column of sediment that is being dated. Uranium/lead (U/Pb) dating is increasingly being applied to Quaternary dune sands, but its age range, while reaching 4.5 billion years ago, begins at one million years ago. Palaeo-magnetism (which has been widely applied to the dating of older loesses) has also been applied to dune sands but is best for the period between hundreds of thousands and millions of years ago. Dating with cosmogenic nuclides is, like thermoluminescence dating (very shortly), a discovery of the last few decades. Cosmogenic dating has been used on dune sands (earlier), but its optimal range is 1 to ~30 million years.

However, the 'datability' of dune sand became one of its greatest strengths as a proxy for climatic change about 30 years ago. Deliverance came in the form of optically stimulated luminescence dating (OSL; and other acronyms for variants). The key discovery, as concerned dunes, was that of Singhvi and colleagues (1982) in Ahmedabad. The major strength of OSL for dune sands is the way in which slip-face accretion quickly and progressively occludes earlier sediment from sunlight, in a much more reliable manner than in most other sedimentary environments (excluding loess). Moreover, OSL can reach back about 500,000 years, well beyond the reach of radiocarbon (Prescott and Robertson, 2008). The greatest problem with OSL is the determination of the level of radiation within a sedimentary particle (the 'dose rate'), which, although measurable, is somewhat dependent on the method of measurement, and anyway is often estimated (Munyikwa, 2005b). When OSL has been compared with dates measured by other techniques, there have been disagreements, but they are now well enough outnumbered by instances of close correspondence of measurements, for there to be fair confidence in the method (Singhvi and Porat, 2008).

Dune-building environments

The determination of a date is, however, only one expectation of the dune archive. Another expectation, which, until the discovery of the applicability of OSL to dune sand, was the main expectation, is evidence of the environment in which the dune was formed, and the archive is much less precise in this role. There are many examples of difficulties in interpreting the environment of dune activity. Were Late Glacial dunes in Upper Michigan reactivated by fires laid by First-Nations huntsmen, fluctuations of the water table or a climatic change (Arbogast and Packman, 2004)?

The reflex interpretation of the discovery of a dune sand is that it signifies a dry period, and this conclusion is often correct, especially where the dune is far from a non-aeolian source of sand. But the one-to-one association of dunes with dry environments may be wrong for two reasons. The first reason applies particularly to dunes close to their alluvial sources of sand, many of which have been found to have been formed in relatively wet rather than dry periods. An example is a climbing dune (Chapter 5) in the canyon of the Niger River (at 13°31'59"N; 02°00'30"E; 400 m), which was built in what is known, from other information, to have been a wet period. The dune sand undoubtedly came from the river, which must have been swollen enough to carry more sandy alluvium than it does today, and there must also have been a season dry enough and long enough for the sand to have been exposed on the river banks, and blown to the dune (Rendell *et al.*, 2003). There are many more examples of dune formation in wet climatic phases (Chapter 9 and Figure 9.1; Clarke and Rendell, 1998; Chase and Thomas, 2007). The second reason for being wary of an activation/rainfall connection is the discovery of the growth of sand ridges in wet-and-windy periods, even in dunes far from their alluvial or littoral sources of sand (shortly).

Pattern as environmental proxy

The use of dune patterns as a proxy for environment is undeniably risky, given the uncertainties about their formation (Chapter 4), and there is also no doubt that there are better proxies. But if the growing portfolio of models of dune pattern (Chapters 5 and 7) were to be focused onto this question, a limited potential to reconstruct ancient environments, and to inform models of dune formation, might be realised. This account looks at the potential of three dune patterns.

Sand ridges
Aside from the analysis of pattern in sand ridges (earlier), the biggest question that their patterns might answer is about the optimal climate for their development. At first sight, the environmental interpretation of the most intensively dated sand ridges (in the south-western Kalahari) is that they were products of dry periods of varying length and severity (Stokes *et al.*, 1997b).

But it has now been claimed that winds in southern Africa may, at times, have been strong enough to create dunes, independently of changes in aridity, and therefore probably in the presence of vegetation (Chase and Brewer, 2009). This 'wet-and-windy' model now has support from two other studies. The first is of dune formation in the Negev in southern Israel (30°56′N; 34°23′E; 5 km), where OSL dating shows that sand ridges were built in two wet-and-windy periods. The first coincided with the first Heinrich Event (when a surge of icebergs suddenly cooled the North Atlantic, during the last glaciation), and the second in the Younger Dryas (later). In the Negev, it is argued, the coincidence of windiness and wetness at the time of dune building supports a model in which vegetation plays an important role in sand-ridge formation (Roskin *et al.*, 2011b). Further support comes from Australia, where there is evidence that sand ridges were formed in moderately humid environments, which, by inference, must have also been windier than they are today (Hesse, 2011). Might the formation of some sand ridges be evidence for wet and windy climates?

Large linear dunes
The best understood large linear dunes are in the Namib (the latest study is Bullard *et al.*, 2011; Chapter 4) and in the Great Sand Sea of Egypt (Besler, 2008;

and later). The large linear dunes in the Algerian sand seas have been known about for much longer than either of these two and have been shown to have great potential for the reconstruction of their environment (later) but little more.

Yet, these three 'known' fields of large linear dunes are atypical (each in its own way) when compared with three large fields of large linear dunes, which are much more regularly spaced, have more consistency of height and are, remarkably, rectilinear. These other sand seas are in the Rub' al Khali, the southern Ténéré, and the Djouf (coordinates in Chapter 4). Apart from a pioneering study of what was then known about the wind environment of the Rub' al Khali (in short, very little; Bagnold, 1951c), some general overviews of Arabian dunes (such as Holm, 1960) and some dating and description of thin sediments deposited in Holocene lakes between the ridges (McClure, 1976, RL), there are no data about the age, morphology or stratigraphy of these dunes. There are even fewer data about the large linear dunes in the Djouf and the southern Ténéré, which, though smaller than the Rub' al Khali, are still huge fields. A list of the other localities of large linear dunes (defined as having a spacing of ~2 km) is given in Chapter 4. Some cover big areas, but many are small, and some do not have the regularity of spacing or the rectilinearity of the three major areas. Many are in now quite humid climates.

Very little is known about the climate associations, or the history of these astonishing landforms, although there has been speculation that linear dunes as a whole may have been the products of accelerated winds in glacial times (Lancaster, 2003, RL). Were their periods of formation short, hot periods of intense winds in which the boundary layer was deep (suggesting a mode linked to the depth of the planetary boundary layer, as discussed in Chapter 4)? If so what does this tell us about the temperature and wind environment of these times? Or were they formed in protracted periods with a very constant pattern of wind bidirectionality (conforming to the model of linear dunes as products of that kind of wind regime)? How does the period of their formation fit with what is known about the repeated swings between extreme aridity and semi-aridity in the Quaternary history of the Sahara and Arabia? Do the presently (apparently) active features share formative environments with the much smaller fields of large linear dunes now in wetter environments?

Parabolic dunes

Parabolic dunes are another pattern with promise as proxy. The present climatic association of these dune forms is very much less ambiguous than that of the other two potential pattern proxies. They occur almost exclusively in what are at present humid climates; and modelling strongly suggests a critical role for plants in their formation (Chapter 6). An association with rainfall of about 300–400 mm yr^{-1} in Argentina (Iriondo *et al.*, 2009) is therefore plausible but is not tested elsewhere. If the need for moisture is accepted, the next question is: why are there so few parabolic dunes in the semi-arid tropics? A map of the dune fields of the Australian interior (Figure 8.1) shows that there are some parabolic dunes in the Wiso dune field on the northern edge of the dry heart of the continent but that there are many more on its south-eastern margins (Hesse, 2010). There are also few (if any) parabolic dunes in the Sahel zone of Sub-Saharan Africa, although there are a (very) few on the pole-ward margins of the African and Arabian deserts (for example, Anton and Vincent, 1986). Given the current models of parabolic dune development (Chapter 6), could the explanation for this distribution be connected with the ecology of the psammophilous plants in these areas?

Second, why are some parabolic dunes wider or longer than others? Some very wide and long parabolic dunes were built in Late Pleistocene Poland and in the Holocene of coastal Queensland. It has been suggested that the degree of elongation of parabolic dunes is related to: wind speed (Gaylord and Dawson, 1987); wind directionality (greater constancy of wind direction being associated with greater length; Pye, 1993b); and sediment supply (Halsey *et al.*, 1990). There is no definitive conclusion.

Finally, why are so many parabolic dunes, both coastal and inland, palaeo-dunes? There have been some observations of near-contemporary formation, but most development has been so slow that it has persisted through several (modest) changes of climate (Chapter 6). These observations aside, does the relic status of so many parabolic dunes mean that winds were stronger in some periods of the Late Pleistocene while still having enough rainfall to support vegetation, and as such contemporary with the sand ridges, the formative environment of some of which is now proposed to have been wet and windy (earlier)? Some parabolic dunes, including some of those on the coast of Queensland, are multiple (up to three or four dunes, of diminishing size, have been built on the same template). There appears to be no alternative to the conclusion that these are relics of a succession of windy periods of diminishing effectiveness.

These questions about inheritance are opened up, although not answered, by two parabolic oddities. The first is the unique 'comb' pattern (discrete groups of up to 20 parabolic dunes), which covers several thousands of square kilometres of the south-western Thar Desert in western India (25°56′N; 71°45′E; 14 km). 'Combs' are ~2 km long with very variable widths of up to ~2 km. The second oddity is the close coexistence of barchans and parabolic dunes in parts of the Canadian Prairies (making oval dunes surrounding a hollow; 50°40′31″N; 109°17′13″W; 2 km). The combined barchan–parabolic pattern occurs also in the Qaidam Desert in China (36°36′44″N; 94°21′21″W; 6 km; Figure 10.9). It has been suggested that both the Canadian and Indian patterns are the outcome of abrupt climatic change. In the Thar, the 'comb' dunes are thought to have formed at a time of high winds and increasing humidity as the south-western monsoon was re-established (between 4000 and 10,000 BP; Singhvi and Kar, 2004). In Saskatchewan, the change is thought to have been much more recent, in the last 200 years (Wolfe and Hugenholtz, 2009). The increasingly successful cellular automaton model of Baas and Nield (2010; Chapter 6) might throw some light on all these questions.

These three patterns all seem to imply that there were much windier periods in the Late Pleistocene than at present, and perhaps also that there were sudden changes in the effectiveness of winds at that time.

The long-term development of sand seas: sediment state

There is more to the history of a sand sea than the variation in wind velocity and directionality (as in Wilson's model), its bodily movement (Chapter 8) or the history of its constituent dunes (the main topic of this chapter). The development of a sand sea, as a whole, is also determined by changing patterns of accretion. This form of development is best understood in the framework of the 'sediment state' concept (Eastwood *et al.*, 2001), which allows more structured discussion about sand seas whose histories are driven by chance external events, and provides a theoretical framework in which to place this reality. It links the long-term history of sediment supply from various sources, with variations in the transport capacity of the wind. The framework can link the consequences of other kinds of interference in sand supply caused by things like the burial and cementation, devegetation and revegetation, and changes in the direction of the wind. Kocurek and Lancaster (1999) illustrated their concept with the history of the Kelso dunes in California (34°54′N; 115°43′W; 11.5 km), whose sand has come mostly from the Mojave River, which at times brought enough sand to build dunes, but at others delivered little, and whose growth was also influenced by climatic changes (Figure 9.1).

Quaternary Dune-Building Climates

Since the start of the Quaternary, 2.6 million years ago, there have been between five and 20 major dry/cold-to-wet/warm cycles (depending on definition). From about 0.8 million years ago, full cycles were ~100,000 years long. Since about 11,000 years ago, the Holocene has been warmer and wetter, and, while still pulsatory, has seen a much lower amplitude of change than the Pleistocene as a whole.

The main driver of these changes has been the varying strength of incoming radiation, itself controlled most strongly by the Earth's orbital rhythms. In very general terms, periods in which there is less incoming radiation are cold and dry; those in which there is more incoming radiation are wet and windy.

But there have been many mismatches between the orbitally forced changes in radiation and changes in geomorphological regimes. There are at least two very different types of mismatch. The first is a delay between a change in the forcing mechanism and the response of geomorphological and ecological systems on the ground. This occurred between 7000 and 8000 years ago in the Mu Us desert of north-central China (located on Figure 10.9), where there was a 10,000–11,000 year lag between a change in solar radiation (the supposed forcing mechanism) and a change to heavier summer monsoonal rainfall, which ended a period of dune and loess formation (Lu Huayu *et al.*, 2005). The second type of mismatch, in contrast, is a sudden change in geomorphological regime, when the change in the forcing mechanism had been gradual. This happened at both the start and the end of the 'African Humid Period' in the southern Sahara in the Mid-Holocene

(Claussen, 2009, RL; shortly). Explanations for these anomalies are constantly being sought, and new models developed.

In general terms, it can be said that climatic changes in the Quaternary and the speed of their onset or ending have both been driven by interactions between changes in the input and spatial pattern of solar radiation, and other processes in the Earth's climatic system (Paillard, 2001, RL). It is possible to assign some local periods of dune formation to different kinds of change brought about by these other processes.

- Starting in the Miocene, continuing but decelerating in the Pliocene, plate tectonics lifted the Tibetan Plateau, which, by excluding the warm, wet south-eastern monsoon, desiccated western China and triggered the onset of dune formation, especially in the Taklamakan. Plate tectonics also created asymmetries in the size and temperature of the Oceans. The Atlantic cooled as it opened up to the Arctic; the Indian Ocean warmed up as it too opened up, but in the tropics. This disparity became one of the drivers of the Thermohaline Circulation, which takes warm water from the Indian Ocean round the Cape of Good Hope, and then twists it northward to the sea off Greenland, where its greater salinity causes it to sink. Most of it then returns to the Indian Ocean. Changes in the strength of this circulation, caused by several processes, reverberate in regional climates. One major disruption of the circulation is thought to have followed a massive discharge of freshwater to the North Atlantic, which brought on the dry and windy Younger Dryas (between about 12,800 and 11,500 years ago, at the end of the Pleistocene), which itself had pronounced impacts on dune formation well beyond the North Atlantic (Broecker, 2006, RL; Chapter 11). Other changes in the circulation may have brought about periods of blowing sand in southern Africa, independently of changes in rainfall (earlier).
- The atmospheric content of CO_2 rose suddenly after the end of the last glaciation, from ~180 ppm to ~280 ppm. Before the change, the global atmosphere could not support as much plant growth as it does now. The resultant thinning of plant cover in the Late Pleistocene has been proposed as part of the explanation for the activity of dunes in the Blue Mountains of eastern Australia and the Nebraska Sand Hills, both during the last major glaciation, when rainfall was not as low as would otherwise have been necessary to allow sand to blow (Hesse et al., 2003a; Mason et al., 2011).
- In the reverse causal direction, vegetation can affect climate: a cover of plants absorbs more radiation than a bare surface. This is Lovelock's 'daisy world' model, applied to the Pleistocene by Claussen (2009). The Early Pliocene desiccation of the Sahara, by denuding its margins of vegetation, may have cooled the whole of Africa and beyond, quite independently of orbitally forced changes (Micheels et al., 2009, RL). The effect may not have been as powerful everywhere: climatic models of the Australian deserts show little reaction to devegetation, and what reaction there was seems to have been more to a change in roughness, than to albedo (Pitman and Hesse, 2007, RL).

- Vegetation on dunes is more sensitive both to desiccation and to increasing rainfall, because sandy soils have freer drainage, greater (plant-) available water capacity, lower losses of water by runoff, and the greater propensity of subsurface horizons to conserve water under evaporative conditions, because of the lower capillary potential of sandy, than finer soils (Weng EnSheng and Luo YiQi, 2008, RL). Thus, dune soils both lose and regain a covering of plants more quickly than do other soils (Ribolzi *et al.*, 2006). These effects, acting on the great swathes of stabilised dunes on the southern margin of the Sahara, and the neighbouring still-active dunes, might have driven the sudden changes that bracketed the 'African Humid Period' (earlier; deMenocal *et al.*, 2000, RL).
- The coincidence of a slight orbitally forced dimming of summer insolation in the northern hemisphere and some large volcanic eruptions in the tropics is what probably brought on the 'Little Ice Age', between ~1350 and ~1850 CE (Wanner *et al.*, 2008, RL), whose effects were most severe in Europe but were detectable as far away as northern Brazil and China (Zhou YaLi *et al.*, 2008). In Europe, the storms of the period increased the rate of coastal erosion, which boosted the supply of sand to beaches and so to coastal dunes (Chapter 11).
- Periodic rises in sea level, as water was released from melting ice sheets, destroyed some coastal dunes, but reinvigorated others by supplying them with more sand from coastal erosion (Chapter 11).
- Concentrations of dust in the atmosphere, which peaked towards the end of most Pleistocene glaciations, may independently have either cooled or warmed the global atmosphere (Overpeck *et al.*, 1996).
- The instability of large ice caps, which may have been the cause of the sudden discharge of cool freshwater into the North Atlantic in the Younger Dryas, and which may then have triggered dune formation in the Negev, and on many coasts (Chapter 11), may also have triggered the Heinrich Events, when coarse sediment was probably rafted on icebergs into the North Atlantic. This also seems to have had an impact on the Negev dunes (later).

There are more of these subsidiary systems, but the effects of most on the activity of dunes were minor or indirect, usually because most were short-lived.

Dunes in the Early- and Mid-Pleistocene

The evidence of dunes in this period is sparse, for several reasons including later recycling (Chapter 9) and burial (most samples known to have come from this period have been retrieved from cores or deep pits, as in Australia; Fujioka *et al.*, 2009).

In Australia, there is debate about the dating of the first Quaternary dunes, but the consensus seems to be that the slide to aridity began ~130,000 years ago in the Mid-Pleistocene (Pillans and Bourman, 2001). There may have been earlier dunes near Lake Amadeus in the Northern Territories (dated by magnetostratigraphy to before 0.98 million years ago, in the Early Pleistocene; Chen XiangYang and Barton, 1991). Analysis of the cosmogenic isotopes [10]Be and [26]Al, in another Australian dune sand, more or less confirms this date (Fujioka *et al.*, 2009).

In North America, sands from the Nebraska Sand Hills and neighbouring dune fields have been dated at 300,000 years ago (Stokes and Swinehart, 1997), giving some confirmation of the evidence of their great age contained in the data about their mineralogical maturity (Chapter 9). Stratigraphically, the oldest dunes in the Gran Desierto of north-western Mexico are from the Middle Pleistocene (Blount and Lancaster, 1990; Figure 8.2). The first dunes in the Great Sand Dunes National Monument in Colorado were being built before 130,000 years ago (Madole *et al.*, 2008).

There are two striking exceptions to the claim that the dunes of this period left no morphological evidence. The first is in south-central Africa (some examples listed in Chapter 4, under 'large linear dunes'; Shaw and Goudie, 2002), where a large area is covered by patterns that are unmistakably those of dunes. Their great age is shown by the incision of rivers into the corridors between some of them, in response to tectonic movement (McFarlane and Eckardt, 2007). OSL dates of windblown sands in this broad area, specifically near the Victoria Falls in Zimbabwe, place them in the Mid- to Late-Pleistocene (Munyikwa *et al.*, 2000). The other main occurrence of ancient dune patterns is an arc north-west of the Great Western Sand Sea in Algeria (earlier). The continuation of the curves of these features by apparently active large linear dunes further south strongly suggests that the silhouettes are those of ancient dunes. They are not as well dated as those in southern Africa, but stratigraphy suggests an Early Pleistocene date (Callot, 1988). Here, as in southern African patterns, great age is suggested by the incision of stream channels in the interdune corridors (31°45′N; 00°19′E; 7 km).

Late Pleistocene Dunes

During the last major glaciation (from ~110,000 to ~10,000 years ago), the wind took geomorphological precedence over a huge proportion of the surface of the globe. The extension of dune-forming conditions may not have been coincident everywhere, but the maximum reach was ~20% of the global land surface, compared with perhaps half that proportion today (Figure 10.5).

The activity of some of these dunes at some time during the last glaciation seriously challenges the imagination. Even more surprising is the remobilisation of some in the Holocene. The most remarkable are in areas where the mean annual rainfall is now over 2 m, as in the Amazon Basin (now under a thick forest canopy; Carneiro Filho *et al.*, 2002), first noticed by Tricart (1974); the Rupununi Savanna in southern Guyana (Teeuw and Rhodes, 2004); central eastern Brazil (Parolin and Stevaux, 2006); and a vast area in the south-western Congo and neighbouring countries, dated by OSL in Zambia and neighbouring areas (O'Connor and Thomas, 1999), which may have been reworked from the earlier Pleistocene dune sands mentioned earlier. In China,

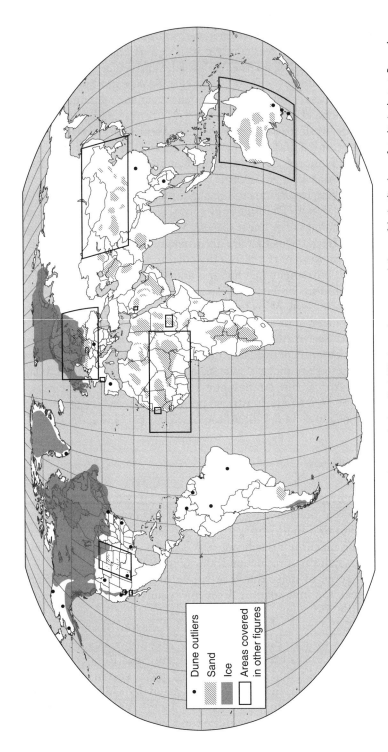

Figure 10.5 Maximum global extent of ice sheets and sand deserts at the Last Glacial Maximum (many sources). Some of the 'outliers' are referred to in the text. Boxes show the locations of other figures.

dunes formed south of the Yangtse, at about 30°N, where the mean annual rainfall is now ~1.5 m (Liu Jian *et al.*, 1997). Many more dunes were built in areas where the mean annual rainfall is now ~1 m: south-eastern Thailand (Sanderson *et al.*, 2001); the Blue Mountains of eastern Australia (earlier); and Wilson's Promontory, the southernmost tip of mainland Australia (Hill and Bowler, 1995). There were also small, isolated, active dune fields in Delaware, the Carolinas, Georgia, Florida, Alabama, Mississippi, coastal Texas, south-western France (Figure 7.1) and Spain (Bateman and Herrero, 2001).

The palaeodunes of the unglaciated far north have been known for much longer than many of the dune fields of the last paragraph, but as products of such environments they still surprise. Large parts of Alaska were not glaciated in the Pleistocene, and some experienced dune-building in the Late Pleistocene, if not earlier. There were, at a maximum, ~30,000 km² of active dunes some time during the last glaciation (Black, 1951). The biggest sand seas were on the north coast, where linear dunes can still be seen, despite their overprinting by permafrost karst lakes (Carter, 1981; 70°19′N; 152°45′W; 30 km). Late Pleistocene windblown sands have also been found on the Arctic shores of continental Canada (Bateman and Murton, 2006). Dunes in other parts of Alaska have had individual histories of activity; some are still active, as are parts of the Great Kobuk dunes (67°03′N; 158°55′W; 10 km; Dijkmans and Koster, 1990). There may be more dune fields to be found in unglaciated Russia than those reported east of Lake Baikal, although limited access to fluvio-glacial sand, and gentle topography, may have restricted the production of dune-ready sand (Shevchenko and Ivanova-Radkevich, 1976).

The main theatres of dune formation in the Late Pleistocene

Much larger dune fields were built in four other parts of the now humid world. In each, dune-building was controlled by a mix of two primary and three secondary processes. The two primary controls were windiness and sand supply; the three secondary controls were aridity, temperature and the atmospheric content of CO_2. Variations in the intensity in these factors brought great variety. Some mixes accelerated dune activity; others slowed it down.

The Great Plains

The biggest of the North American palaeo-sand seas is the Nebraska Sand Hills (50,760 km²; 41°N; 101°W; 200 km). There are many smaller now-stabilised sand seas elsewhere on the High Plains (Figure 10.6). Dating has shown a long history of alternating activity and stabilisation (Rich and Stokes, 2011).

The dunes on the Plains were driven by strong north-westerly winds, a pattern that has been confirmed by climate models (Kutzbach and Wright, 1985, RL). The westerlies of the time were constricted by the ice sheets and the Rockies, and thus accelerated and redirected round the ice sheet. Dunes were probably built with sand inherited from earlier Pleistocene arid periods, as

Figure 10.6   Palaeo-dunes on the High Plains of the USA (Muhs and Holliday, 2001).

indicated by the great maturity of the sand in the Nebraska Sand Hills (Chapter 9; Figure 9.3 and earlier). The Late Pleistocene on the Plains was drier and colder both than it is today and than Europe in the Late Pleistocene. One piece of evidence of this is the limited extent of parabolic dunes on the Plains, compared with their prominence in dune patterns in Europe (later). Lower levels of atmospheric CO_2 during the last glaciation, may also have restricted plant growth (earlier).

Europe

Most of the active dunes of the period were within the 'European Sand Belt', which stretches from eastern England to European Russia (Figure 10.7). Smaller dune fields developed in many satellite areas, as in Hungary (Ujhazy *et al.*, 2003).

The main distinction between the dune-friendly climate of Europe and the other two temperate theatres of dune formation in the Late Pleistocene, was the greater humidity in Europe, the evidence for which is both palaeobiological and

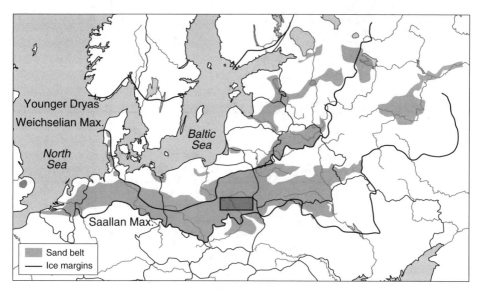

Figure 10.7   European Sand Belt, from Zeeberg (1998, figure 1, p. 128). The coincidence of the limit of the Saalian ice and the Sand Belt is an indicator of the importance of glacial outwash sands to dune formation. The Saalian was a Middle Pleistocene Ice Age; 'Weischellian' refers to the limit of the ice in the last Ice Age; 'Younger Dryas' refers to the limit of a Holocene glacial re-advance. The box shows the position of Figure 10.8. Reprinted with permission from John Wiley & Sons.

geomorphological. The geomorphological evidence lies first in the dominance of 'coversands' (Chapter 4) in the western Sand Belt. The discovery that some of the western coversands have a topography of subdued parabolic dunes does not conflict with this conclusion, given the climatic associations of parabolic dunes (Jungerius and Riksen, 2010; Figure 4.15; earlier). The eastern Sand Belt was drier than the western, but the frequency of parabolic dunes (and, less so, of coversands) suggests that here too, conditions were wetter than on the Great Plains (Figure 10.8). Some of the eastern dunes are, even today, 30 m high, and some travelled 100 km from their origins before they were stabilised (Högbom, 1923b; Manikowska, 1995).

A second major difference between the European dunes in the Late Pleistocene and those in the other dune-forming theatres was that the European dunes drew their sand primarily from fluvio-glacial and dune sand that had already been recycled several times during the Pleistocene. The most striking evidence of this inheritance is the coincidence of the border of the Sand Belt and the limit of the Saalian Ice (an earlier glaciation) (Figure 10.7). In Europe, winds during the Last Glaciation were stronger than they are today (Renssen *et al.*, 2007) but may never have reached the power of the winds on the Great Plains or in China.

Figure 10.8 Stabilised Late Pleistocene parabolic dunes in Poland. The 'deflation area' provided the sand for the dunes (Högbom, 1923). Located on Figure 10.7. Reprinted with permission from John Wiley & Sons.

North-eastern China

The Chinese theatre of Late Pleistocene dunes was probably the windiest of the temperate dune theatres. The strength of winds in the Late Pleistocene is indicated by the rate of accumulation of dust in deep-sea sediments in the North Pacific, which plummeted from >500 mg cm^{-2} kyr^{-1} during the last glaciation to ~120 mg cm^{-2} kyr^{-1} today (Hovan et al., 1991). The Chinese dunes also had access to much more generous supplies of sand than did the other two temperate theatres, partly a consequence of greater tectonic activity and more accentuated relief. It is likely that Late Pleistocene dunes, here as in the other theatres, recycled sand from earlier Pleistocene dunes (Li Xiaoze and Dong Guangrong, 1998).

The best researched of the now-stabilised Chinese dune fields is the Mu Us, within the great bend of the Huang He (Figure 10.9). Dunes in the Mu Us were repeatedly reactivated and stabilised in the Late Pleistocene. Dune formation (in windy periods) alternated with the deposition of loess (in calmer and wetter interludes). In some of the windier periods, some Mu Us dunes invaded eastward through topographic corridors (38°23′04″N; 109°58′44″E; 60 km; He Zhong et al., 2010b).

The temperate southern hemisphere

The biggest Late Pleistocene dune field in Argentina is on the east-central Pampas. It is an arc of large linear dunes, now with subdued topography, stabilised and extensively cultivated (35°48′S; 62°W; 75 km; Zárate and Blasi, 1993). Dunes were also active at this time elsewhere in Argentina, high above the Great Escarpment in South Africa (Marker and Holmes, 1993), in extreme north-eastern Tasmania (40°47′S; 148°02′E; 8 km; east of Bridport; Duller and Augustinus, 2006), at Wilson's Promontory in Victoria and in the Shoalhaven basin in south-eastern New South Wales (Nott and Price, 1991) (all the Australian locations named in the text are shown in Figure 8.1).

The semi-arid tropics

The existence of dunes of Late Pleistocene age in the Sahel of northern Africa has been recognised for some time (Tricart and Brochu, 1955; Grove and Warren, 1968; Warren, 1970; Figures 10.10 and 10.11). Tricart and Brochu's field area was the sand seas of Trarza and Cayor (16°20′N; 16°00′W; 97 km, located on Figure 10.11). These studies were able to outline the extent of the dunes, and to infer something of the sequence of dune formation, but little more. Even now, few of these sands have been dated by OSL. In Burkina Faso, OSL has shown that dunes were active in three Late Pleistocene periods: ~40,000, ~17,000 and ~14,000 years ago, more or less confirmed by another OSL study of dunes in north-eastern Nigeria (coordinates earlier; Albert et al., 1997; Stokes and Horrocks, 1998).

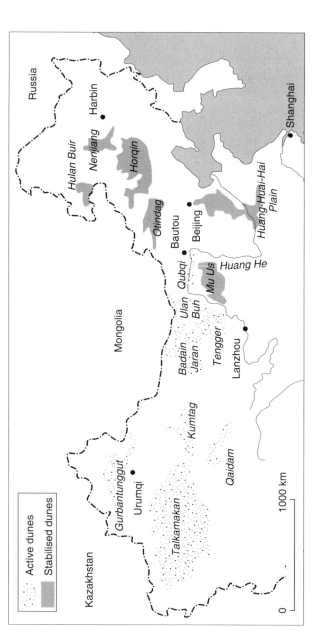

Figure 10.9 Stabilised and active sand seas in China (many sources).

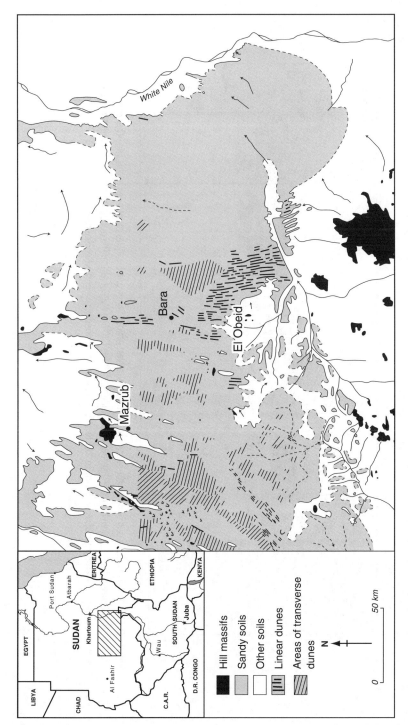

Figure 10.10 Extent of stabilised dunes (in Arabic, '*qoz*') and their patterns in the central Sudan (simplified after Warren, 1968b).

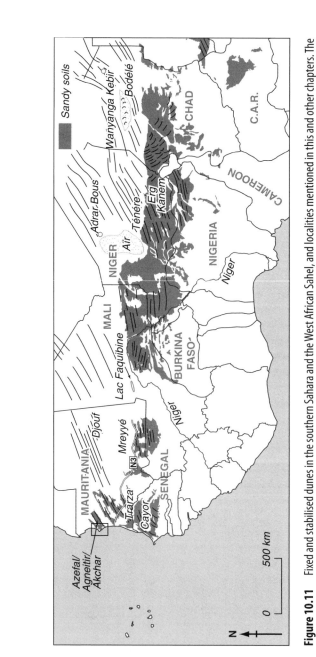

Figure 10.11 Fixed and stabilised dunes in the southern Sahara and the West African Sahel, and localities mentioned in this and other chapters. The extent of the stabilised dunes is taken from the distribution of sandy soils on the FAO Soil Map of Africa. The dune patterns are diagrammatic; some are taken from Grove and Warren (1968). Also shown are the locations of places mentioned in this and other chapters.

The conditions that favoured dune building on the south side of the Sahara are as poorly understood as their dates. There is little, if any, evidence for the inheritance of sand from earlier dune-building periods (in contrast to the emerging evidence in most of the other main theatres of dune-building), although this does not mean that there was none. Very tenuous evidence for a great age is the speculation that they have accumulated at the end of long active trajectories of wind-laden sands, which sweep south-westward over the southern Sahara into the Sahel and Sudan (Figures 10.10 and 10.11; El-Baz et al., 2000).

The stabilised dune systems of southern Africa, like those in the Sahel zone, have been known for some time (Grove, 1969) but had to wait for a spurt of OSL dating in the last 20 years before their importance could be established. This work has been extensively discussed earlier in this chapter (for example, Thomas and Wiggs, 2008).

In the Thar Desert in western India, near the research centres at which OSL was found to be feasible for windblown sands, OSL dates are now thick on the ground. They give a diverse picture (perhaps just because of their quantity). Some support the proposal that there was only one Late Pleistocene period of dune activation; others claim up to four (Andrews et al., 1998; Juyal et al., 2003a). The close of the (perhaps each) episode of dune formation was time-transgressive, moving south-westward, as in the Holocene in northern Europe and North America (later).

In central Australia, a compilation of OSL ages from the Lake Eyre basin shows that dune formation intensified after 40,000 years ago, during the Last Glaciation (Hesse et al., 2004, RL). In some places, dune formation was driven as much by new supplies of alluvium, perhaps stimulated more by higher rainfall in the basins of the rivers draining south-westward from wetter areas, than by aridity (Cohen et al., 2010; Figure 9.2). A long-lasting dwindling of the supply of sand, as rivers dried up, may have later constrained dune building (Hesse, 2010). Many of the Australian dunes, especially those on the margins of the desert, where palaeosols have developed, hold evidence of repeated phases of activity throughout the period of their formation. The last episode of widespread sand movement left deep deposits of sand on many dune crests, since when there has been little activity. In the Murray-Darling basin of south-eastern Australia, the now-stabilised (and cultivated) linear dunes of the Mallee were active from 25,000 to 15,000 years ago (35°15′S; 140°24′E; 22 km; Pell et al., 2000). Throughout the Australian dune fields, most dunes yield dates from the Late Pleistocene or later, almost certainly because recycling has destroyed the datable signals of earlier sands (Figure 9.3).

The present deserts

Despite the mobility of dunes, there are some datable sands in contemporary deserts: on the stable plinths of linear dunes (Chapter 4); in soils (though rare); in forced dunes (earlier); and from cores.

The most detailed histories of desert dunes are from North America. The Algodones dunes, the biggest still active dune field in the USA, were built in two phases: an older one of large transverse dunes, and a later one of small transverse and barchan dunes (Derickson *et al.*, 2008). The earliest sands were laid down only 31,000 years ago (Stokes *et al.*, 1997a). The Kelso Dunes in the central Mojave are a much smaller dune field (34°54′N; 115°43′W; 11.5 km); most of the others (many in the same sand supply chain as Kelso) are even smaller (Figure 9.1). All these dune fields began to accumulate in a semi-arid climatic phase in the Late Pleistocene, when rivers and lakes were delivering larger amounts of sand than in fully arid phases but were not fully reactivated until the next arid phase (Clarke and Rendell, 1998).

The best data in the Sahara are for the Azefal/Agneitir/Akchar dune fields near the Atlantic coast of Mauritania (Figure 10.12, earlier; 19°15′N; 16°W; 40 km,

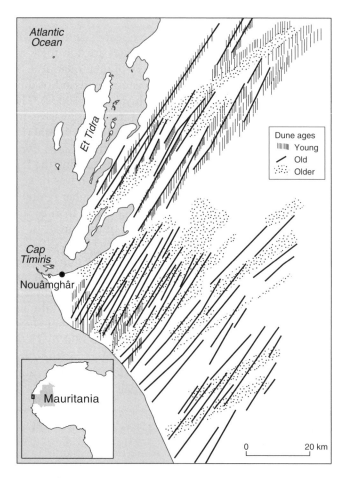

Figure 10.12 Patterns of palaeodunes in the Azefal/Agneitir/Akchar area on the Saharan boundary in western Mauritania (Lancaster *et al.*, 2002). Dunes of successive ages are lower and have closer spacing and different orientations, a result of shorter periods of formation and different wind systems. Located on Figure 10.10.

also located in context on Figure 10.11), which is on the border between active and stabilised dunes. There are three populations of dunes here: the oldest are the most widely spaced; their slopes are the most degraded, and the colour of their sand is the reddest; their orientation is north-easterly. OSL shows their age to be between 25,000 and 15,000 years old (the Late Pleistocene). The remainder of the series, which was formed in the Holocene, is discussed shortly.

Another good set of numerical dates comes from the Great Sand Sea of Egypt. There were two periods of dune formation in the Late Pleistocene, and, less certainly, one earlier and two later. The only evidence for the oldest, which probably dates from the Early Pleistocene, is from sand taken from a deep core. The dunes in the main body of the sand sea date to 23,000 years ago (during the last glaciation). They are dominated by widely spaced north–south linear dunes (Bagnold's 'longitudinal dunes'; 1941, pp. 222–229; 25°21′N; 26°50′E; 80 km), which seem to have been formed by northerly winds, the direction from which most winds still blow. These dunes were degraded in a later wet period, before the formation of a Holocene series (28°14′N; 25°31′E; 30 km). These dunes are large, although smaller than the first series. They are also more closely spaced, more sinuous and more asymmetric than the first series, and their steeper slopes face east. They appear to have been built by transverse to strong westerly winds, between 20,000 and 7000 years ago, straddling the Pleistocene–Holocene boundary (Besler, 2008).

In the Taklamakan, dunes were inactive during glacial maxima, perhaps because of frozen and snow-covered conditions, but as sediment was released by glacial meltwaters and as snow cover declined in the Late Pleistocene, the wind was again able to move sand (Wang Yue and Dong Guangrun, 1994). In the body of the sand sea, conditions swung between windier and less windy, while the climate remained arid throughout, so that dune formation continued regardless, albeit at different rates (Feng Qi and Liu Wei, 2005). Further east, the Badain Jaran Desert (in which dunes are now active) experienced marked and frequent humid periods throughout the Late Pleistocene (40°09′N;101°58′E; 190 km; Yang Xiaoping, 2004).

Dunes in the Holocene

In most parts of the world, the mid Holocene (the 'Climatic Optimum') was warmer and wetter than both the earlier and later. Another general climatic trend in the Holocene was the slow weakening of solar radiation in the northern hemisphere, which gradually diminished the strength of the West African monsoon, and therefore led to the gradual desiccation of the area over which it blows (Wanner et al., 2008, RL). Superimposed on longer-term changes like this were events that had regional rather than global impacts. The most prominent of these were the Younger Dryas, and the 'African Humid Period' in West Africa (earlier), the second of which, perversely, coincided with a strong drought on the Great Plains (Shin Sang-Ik et al., 2006, RL). As well as that event, there were many less severe and less pervasive droughts, as on the Great Plains (later).

These episodes aside, the main feature of the dune archive for the Holocene has been great variability, in space and time. One part of the explanation may be because the very superabundance of Holocene dune sand, recycled from older sand (Chapter 9), has left a legacy of easily reactivated dune sand, which is both more sensitive to weak climatic fluctuations than the older periods and more accessible, given its survival in greater quantities. Thus, quite subtle changes in climate can be detected, as in the Canadian Prairies, where decadal-scale changes have been detected (Hugenholtz and Wolfe, 2005b).

Another cause for variability was the growing number of people, now equipped to burn, graze, cultivate, fight, flee and hide, all of them easier on the open, more tillable soils, better grazing, greater defensibility and easier travelling on the grasslands on now-stabilised dunes, than to the woods, swamps and heavier soils in the rest of the landscape. The activities of people undoubtedly accelerated the reactivation of dunes, although mostly in small patches. Stabilised dunes near Krakow in Poland had already been inhabited during the last glaciation, but people did not make a serious impact there until the Neolithic, which began in Poland ~6000 years ago. At that time, the acceleration of sand movement, the redeposition of sand (as in ponds) and dune movement have been attributed to people rather than to climate (Dulias *et al.*, 2008).

In Europe, human intervention cut both ways: towards mobilisation or stabilisation. Agriculture on the easily tilled sandy soils survived, and for long periods, by using a practice that diminished rather than increased wind erosion. This was the transfer of litter and turf from uncultivated land to fields, with the intention of improving the fertility of sandy soils (Mikkelsen *et al.*, 2007). But whether for cultural or environmental reasons (wars, disease, etc.) or because of drought, this practice was at some times and in some places abandoned, and dunes, some big enough to bury settlements, reappeared (Castel *et al.*, 1989; Mauz *et al.*, 2005).

Accusations of environmental recklessness are more often aimed at pastoralists than at cultivators, but their record is as equivocal. Reindeer herders probably did disturb dune grasslands in northern Finland (Käyhkö *et al.*, 1999b), but few such claims are as secure. For example, pastoralists arrived on the Tibetan Plateau 6000 years ago, in a warm phase, but by the time their numbers had grown to the point when they might have had a detectable impact, it is difficult to distinguish their role from the signals of climatic cooling (Schlütz and Lehmkuhl, 2009). Other pastoralists have used the sandy soils of the Negev of southern Israel and the nearby areas of Sinai for at least 4000 years, and here as elsewhere, there were fluctuations in climate in the Holocene. Distinguishing the effects of the climatic fluctuations from those of pastoralism, on the mobilisation of windblown sand, is also not simple (Warren and Harrison, 1984, RL; Tsoar and Goodfriend, 1994).

The deglaciated North

As European and North American ice sheets retreated, they left a patchwork of small, sandy lake deltas, kames and eskers, which stretched from the southern limits of the ice in both Poland (Zeeberg, 1998), and southern Michigan to the Arctic Ocean. In aggregate, these dunes cover a great area, for example 12,000 km² in Eastern Upper Michigan alone, but most, there and elsewhere, are in small patches (Loope et al., 2012). When some of these sites emerged from the ice, dunes were built by winds blowing clockwise round the ice sheet, which in some cases was a diametrically opposite direction to winds that formed later dunes (Filion, 1984). On emergence from the ice, these and other sites were subject to two time-transgressive northward-moving waves. The first was a rapid mobilisation of the sand by the wind and its collection into dunes, as on an esker near Rokuanvaara in Finland (64°32'46"N; 26°36'41"E; 10 km; Aartolahti, 1973). The second wave was the smothering of windblown activity by a vegetation succession (to woodland in the south, to tundra in the north, also as on the esker at Rokuanvaara). Small patches of these stabilised dunes were later reactivated but quickly restabilised. On the drier and colder North American sites, as in Saskatchewan, there were more phases of reactivation and stabilisation, triggered both by minor climatic changes and by people (Hugenholtz et al., 2010). The biggest area of still active sands in the Arctic (of both continents) is the Athabasca dune field in northern Canada (59°00'N; 109°17'W; 30 km).

The mid-latitudes

Further south, the Holocene experienced many minor changes in climate. On the drier, more westerly Great Plains, most lunettes, for example, were reactivated in the Holocene (Bowen and Johnson, 2008, RL). On the Great Plains of North America more generally, aridity and much colder winters (now, and even more so in parts of the Holocene) than in the Europe of the time positioned dunes on the High Plains much closer to thresholds of reactivation, whether in response to local or regional triggers. Regionwide droughts caused many dune fields to cross a threshold at the same time. One of these was a major event in the Mid-Holocene, at a peak in orbitally driven solar radiation, perhaps in conjunction with a dry El Niña phase (Shin Sang-Ik et al., 2006, RL). A patchier pattern of reactivation characterised the period between 4100 and 4300 years ago and continued in historical times, but a measure of coordinated reactivation returned the 'Medieval Climate Anomaly', 900 CE to 300 CE, and at times in the 19th century (Forman et al., 2006; Miao Xiaodong et al., 2007b; Hanson et al., 2010). The most damaging dry periods, like the episode just mentioned, coincided with the cool part of the ENSO cycle in the eastern Pacific (Menking and Anderson, 2003). Some activations were time-transgressive, beginning with the dunes nearest to the thresholds of movement in the dry west, and moving eastward to less stressed dunes (Williams et al., 2010). Stabilisation was time-transgressive in the other direction.

In Europe, the timing of the formation of coversands is disputed. Some authorities see many periods of reactivation; others see only one, but with a patchy spatial distribution (Kolstrup, 2007a). In Poland, dune activity in the Holocene became more intermittent, and any new dunes were distinctly smaller than those of the Late Pleistocene, a function of shorter periods of dune formation, of weaker winds, of returning vegetation or of restrictions in sand supply (Gozdzik, 2007).

In north-eastern China, dune activity spilled over from the Late Pleistocene into the Holocene, but most dunes had been stabilised by the time of the Holocene Optimum (between ~9000 and ~6000 years ago). This was followed by one severe and one or more periods of less severe reactivation, perhaps ending with a phase of activation caused by disturbance. There may be some convergence in the results of these studies, but there are also hints of a patchy pattern, as in western Europe and on the Great Plains (compare Lu Huayu *et al.*, 2005; Sun Jimin *et al.*, 2006; Zhao Hua *et al.*, 2007).

The semi-arid tropics

There is ample evidence of Holocene activity in the Sahel, where some dramatic shifts in rainfall have been dated numerically, as in the Mema in Mali (Makaske *et al.*, 2007). There was also Holocene reactivation on the north side of the Sahara in western Libya (Giraudi, 2005) and in eastern Kenya (Mahaney and Spence, 1986). Some of these phases may have been short, and may have started and ended suddenly, as in Mauritania (Hanebuth and Henrich, 2009). In the Thar Desert in western India, studies based on OSL dating have found up to five reactivations in the Holocene (Thomas *et al.*, 1999a).

In the southern hemisphere, no Holocene reactivation has been found in the northern and eastern Kalahari, but dunes in the drier south-west may have been patchily active for most of the last 120,000 years (Stone and Thomas, 2008). Peak activity was at about the Pleistocene–Holocene turnover (Telfer and Thomas, 2007). Many of the Australian dunes, especially those on the margins of the desert, where palaeosols have developed, have evidence of repeated phases of activity in the Holocene. As in so many other parts of the world, Holocene reactivation was very patchy (Fitzsimmons *et al.*, 2007a).

The present deserts

The most detailed histories of desert dunes are from North America, perhaps because of their accessibility. The high proportion of Holocene dates and the creation of many dunes in moderately humid phases have already been discussed. Except for an underlying sand (earlier), the Algodones dunes were built in two phases, both in the Holocene: an older one of large transverse dunes, and a later one of small transverse and barchan dunes (Stokes *et al.*, 1997a). The story at White Sands in New Mexico is rather different. For a start, the main phase began only ~7000 years ago. The critical difference was that dune growth was driven by the availability of gypsum sand when lake levels fell, rather than by dry or windy periods (Kocurek *et al.*, 2007).

Given the present high rate of sand and dune movement and the much bigger dunes and dune fields of the presently extreme deserts in northern and southern Africa, it is not surprising to find that they were active in the Holocene. For example, a large dune in the Namib was reworked within the last 5700 years and moved 300 m within the last 2500 years (Bristow *et al.*, 2007a). The Mauritanian dunes, whose first stage was described earlier, continued to develop in the Holocene (Figure 10.12). Unlike the Namib, the Mauritanian area experienced distinct changes in the direction and perhaps the strength of winds at this time. The second set of the Mauritanian dunes are closer together than the first, probably because of a shorter period of activity, which was between 10,000 and 13,000 years ago, thus coinciding with the Younger Dryas (earlier). The third and most recent set of dunes in Mauritania were formed in the last 5000 years. They are the most closely spaced and have an orientation directly north. The orientations of all three periods of dune development in Mauritania are conformable with models of winds in the Late Pleistocene, and with data from other onshore and some offshore sediments (Lancaster *et al.*, 2002a). The Egyptian sequence of the Great Sand Sea (earlier) was also completed in the Holocene, when a bimodal wind regime created seif dunes (Chapter 4) on the summits of both generations of older dune (lighter-coloured sand on the Google Earth image, whose coordinates were given earlier) (Besler, 2008).

In eastern Arabia, most of the dunes built in the Holocene were small and developed on the crests of earlier dunes (Lancaster *et al.*, 2003, RL), as in Egypt. However, some large dunes were also formed, particularly the biggest dune in the area. This is the 150 m high crescentic dune that overlooks the oasis of Liwa from the north (23°07′N; 53°47′E; 100 km), which grew almost wholly within the Holocene. A spurt of growth between 3000 and 2000 years ago was driven by a large delivery of sand from an increase in marine erosion (remobilising older dune sand) as sea level rose after the end of the last glaciation (Stokes and Bray, 2005; as in the southeastern Wahiba Sands earlier). Dune building in the Holocene also occurred further north in the UAE, in a wetter (rather than wet) environment, where vertical growth of ~3.3 m in a thousand years occurred about 10,000 years ago, perhaps partly because of disturbance by domestic grazing animals (Goudie *et al.*, 2000).

References

Bowen, M.W. and Johnson, W.C. (2008) 'Playa-lunette systems and paleoenvironmental change on the central High Plains', *Abstracts with Programs Geological Society of America* 40 (6): 147.

Broecker, W.S. (2006) 'Was the Younger Dryas triggered by a flood?', *Science* 312 (5777): 1146–1148.

Claussen, M. (2009) 'Late Quaternary vegetation-climate feedbacks', *Climate of the Past* 5 (2): 203–216.

Dabbagh, A.E., Al-Hinai, K.G. and Khan, M.A. (1998) 'Evaluation of the Shuttle Imaging Radar (SIR-C/X-SAR) data for mapping paleo-drainage systems in the Kingdom of

Saudi Arabia', in A.S. Alsharhan, K.W. Glennie, G.L. Whittle and C.G.S.C. Kendall (eds.), *Quaternary Deserts and Climatic Change, IGCP 349*, December 1995, Al Ain, United Arab Emirates. Rotterdam: A.A. Balkema, pp. 483–493.

deMenocal, P., Ortiz, J., Guilderson, T., Adkins, J., Sarnthein, M., Baker, L. and Yarusinsky, M. (2000) 'Abrupt onset and termination of the African Humid Period: rapid climate responses to gradual insolation forcing', *Quaternary Science Reviews* 19 (1–5): 347–361.

Drake, N.A., El-Hawat, A.S., Turner, P., Armitage, S.J., Salem, M.J., White, K.H. and McLaren, S. (2008) 'Palaeohydrology of the Fazzan Basin and surrounding regions: The last 7 million years', *Palaeogeography Palaeoclimatology Palaeoecology* 263 (3–4): 131–145.

Hesse, P.P., Magee, J.W. and van der Kaars, S. (2004) 'Late Quaternary climates of the Australian arid zone: a review', *Quaternary International* 118–119: 87–102.

Kutzbach, J.E. and Wright, H.E. Jr. (1985) 'Simulation of the climate of 18,000 years BP: Results for the North American/North Atlantic/European sector and comparison with the geologic record of North America Original Research Article', *Quaternary Science Reviews* 4 (3): 147–187.

Lancaster, N. (2003) 'Linear dunes; are they a product of glacial wind regimes?', *Congress of the International Union for Quaternary Research* 16: 128.

Lancaster, N., Singhvi, A., Teller, J.T., Glennie, K. and Pandey V.P. (2003) 'Eolian chronology and paleowind vectors in the northern Rub al Khali, United Arab Emirates', *Congress of the International Union for Quaternary Research* 16: 141.

McCauley, J.F., Breed, C.S. and Schaber, G.G. (1997) *The Sahara Paleodrainages: SIR-C/X-SAR Flights*; Final Report to NASA (Jet Propulsion Laboratory).

McClure, H.A. (1976) 'Radiocarbon chronology of late-Quaternary lakes in the Arabian desert', *Nature* 263 (5580): 755–756.

Micheels, A., Eronen, J. and Mosbrugger, V. (2009) 'The late Miocene climate response to a modern Sahara desert', *Global and Planetary Change* 67 (3–4): 193–204.

Paillard, D. (2001) 'Glacial cycles: Toward a new paradigm', *Reviews of Geophysics* 39 (3): 325–346.

Pitman, A.J. and Hesse, P.P. (2007) 'The significance of large-scale land cover change on the Australian palaeomonsoon', *Quaternary Science Reviews* 26 (1–2): 189–200.

Shin Sang-Ik, Sardeshmukh Prashant, D., Webb, R.S., Oglesby, R.J. and Barsugli, J.J. (2006) 'Understanding the mid-Holocene climate', *Journal of Climate* 19 (12): 2801–2817.

Wanner, H., Beer, J., Butikofer, J., Crowley, T.J., Cubasch, U., Fluckiger, J., Goosse, H., Grosjean, M., Joos, F., Kaplan, J.O., Kuttel, M., Muller, S.A., Prentice, I.C., Solomina, O., Stocker, T.F., Tarasov, P., Wagner, M. and Widmann M. (2008) 'Mid- to Late Holocene climate change: an overview', *Quaternary Science Reviews* 27 (19–20): 1791–1828.

Warren, A. and Harrison, C.M. (1984) 'People and the ecosystem: biogeography as a study of ecology and culture', *Geoforum* 15(3): 365–381.

Weng EnSheng and Luo YiQi (2008) 'Soil hydrological properties regulate grassland ecosystem responses to multifactor global change: a modelling analysis', *Journal of Geophysical Research* 113: GO3003.

Chapter Eleven
A History of Coastal Dunes

Changing sea level is the strongest rhythm of formation and destruction in coastal dunes, although they do not wholly avoid the influence of the climatic rhythms that drive inland dunes, particularly episodes with stronger winds. The same applies to coastal calcareous aeolianites, although they better resist destruction, given their lithification.

Long Sequences

Though rare, some ancient coastal dunes do survive. The most remarkable sequence occurs in south-eastern Australia. It has been preserved by the slow and steady uplift of the south-eastern quadrant of the continent, as it is nudged northward by plate tectonics. The full sequence extends 500 km inland and dates back 1.6 million years (Figure 8.1; Bowler et al., 2006). The Late Quaternary dunes, close to the present coast in the Coorong of South Australia, are calcarenites (shortly; 36°31′S; 139°51′E; 4 km). At one site on this coast, one new dune ridge has been created every 80 years for the last 3900 years (Murray-Wallace et al., 2002).

Sea Level

After the last major glacial meltdown, sea levels rose by about 130 m between 15,000 and 8000 years ago. In the 'Cooper–Thom' model, the waves of rising seas cut into coastal exposures, and created new dunes with the sand they released

Dunes: Dynamics, Morphology, History, First Edition. Andrew Warren.
© 2013 John Wiley & Sons, Ltd. Published 2013 by John Wiley & Sons, Ltd.

(Pye, 1984b). The model has been quantified for coastal dunes on Groote Eylandt in the far north of Australia (13°55'S; 136°46E'; 11 km; located on Figure 8.1): when the sea level was at 40 m below the present level, dunes were not being built; as sea levels rose from −40 to −5 m, the dune system grew, fed by the sand released by marine erosion; from −5 m to +5 m, the dune system lost sand to the wind; a rise to +5 m destroyed the dunes (Shulmeister et al., 1993).

The model applies better to gently than to steeply shelving coastal waters. For example, it seems to apply to the gently shelving North Sea coast of the Netherlands, where the 'Younger Dunes' were built after a slight rise in sea level between AD 800 and 1650 (Arens and Wiersma, 1994a). The model also applies where a rise in sea level is caused by tectonics, as in the gently shelving coasts of south-eastern England, which have been slowly sinking for many millennia, and experienced almost continuous Late Pleistocene dune formation (Orford et al., 2000). The gently sloping coasts of lakes respond in much the same way, as on Lake Michigan over the last 5000 or 6000 years (43°07'N; 86°16'W; 210 m; Hansen et al., 2004). The world's oceans are now gently rising, and beaches and dunes are being destroyed almost everywhere. The consequences are discussed in Chapter 14.

Because sea levels have moved up and down repeatedly over the last few thousand years, individual histories of the supply of sand to coastal dunes are much more complex than this outline. In northern Ireland, the sediment left by retreating glaciers at the end of the last glacial period was raked over and sorted by the waves of falling seas, as rapid isostatic uplift of the land followed the melting of the glaciers. The waves of the following period of eustatically driven rising seas then drove onshore the sand produced by the first episode. The result was generous amounts of potential dune sand on the benches that were uncovered when the sea level fell again in the Holocene (Carter and Wilson, 1993).

Other Controls

The Cooper–Thom model does not explain the history of all coastal dunes. In many areas, the supply of sand to dunes is independent of a rise in sea level; in other words, it comes from sources other than the erosion of coasts. This is said to have been the case in south-eastern Australia, where strong winds in the postglacial both invigorated long-shore drift and accelerated onshore movement of sand across beaches (Lees, 2006). In Denmark, and elsewhere on European coasts, the rate of onshore sand movement in the windy Little Ice Age, between the 16th and the 19th centuries CE, has been calculated to be two to three times greater than it is today (Aagaard et al., 2007; Clarke and Rendell, 2011). The Little Ice Age may even have left a similar mark on coastal dunes in southern Brazil (Sawakuchi et al., 2008). There were as many as five of these stormy periods in the Holocene of north-western Europe, none as intense as the Little Ice Age, most of which saw the growth of coastal dunes (Clemmensen et al., 2009).

Less intense and shorter-term periods of storminess also affect the activity of coastal dunes. The North Atlantic Oscillation (NAO) (Chapter 10) affects storminess at a quasi-decadal rhythm, which accelerates or decelerates the activity of dunes on the Portuguese coast (Clarke and Rendell, 2006). Further south, on the north-eastern Brazilian coast, it is the activity of the Southern Oscillation (the equivalent in the southern hemisphere of the NAO) that supplies the main rhythm of activity (Figure 11.1).

There are yet more ways in which the Cooper–Thom model has been flouted. Where there is little vegetation, sand exposed by a retreat in sea level can be blown far inland, well beyond the coast, as on the coasts of the Arab/Persian Gulf and Oman (shortly). Temporal changes in aridity can also, and alone, initiate coastal dune formation, as on the western coast of South Africa (Chase and Thomas, 2006). In northern Ireland, yet another explanation for dune formation, quite independent of sea level, may have been the unavailability of 'accommodation space', which is a function of the steepness of the immediate coast, and of changes in sea level, which may either open up a space to dune formation that had previously been inundated or inundate one and thus preclude dune formation (Wilson and Braley, 1997).

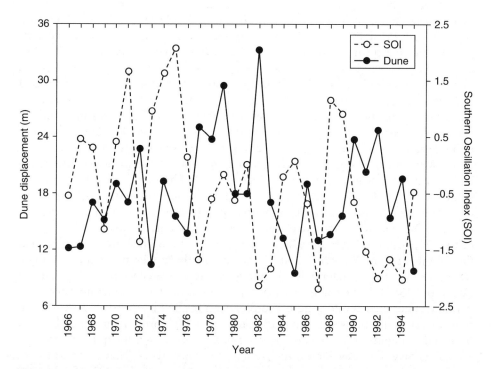

Figure 11.1   Displacement of coastal dunes in north-eastern Brazil versus an index of climatic variability (from Maia *et al.*, 2005). With kind permission from *Journal of Coastal Research*.

Calcareous Aeolianite

Calcareous dune sands are introduced here for two reasons: most are held in palaeodunes, and most are coastal. Most carbonate dune sand is derived from the exoskeletons of a great assortment of marine organisms. Hence, one of the drivers of the supply of calcareous sand is marine biological productivity, which is greatest where: (1) there is little input of fluvial or cliff-derived sediment, which would diminish marine productivity (thus carbonate and non-carbonate dunes seldom coexist); (2) there are shallow offshore shelves, on which these organisms thrive (this form of control may override others in Bermuda and the Bahamas, where at one stage, great quantities of calcareous sand were liberated not so much by increased wave action as sea levels rose as by intensified biological activity following the inundation offshore shelves; Carew and Mylroie, 2001); (3) the climate is warm enough to encourage high marine productivity, as seen in the abundance of coastal calcareous aeolianite between 40° north or south (Brooke, 2001); (4) perhaps there is marine upwelling; and finally (5) waves have sufficient power to break down skeletal remains to particles of sand size.

Although calcite (which forms the bulk of carbonate sands) is slightly denser than quartz, shell fragments are very porous, and most are more platy in shape than quartz sands, both of which properties allow readier movement by the wind (Figure 1.10). Miliolite sands, derived from species of the miliola genus, for example, are blown 800 km inland in western India (Goudie and Sperling, 1977), but most calcareous sands, there and elsewhere, have shorter journeys for two reasons: they are broken down in saltation; and, once in a dune, they are quickly immobilised by lithification (shortly). The great bulk of the Indian miliolites are coastal, as in Kathiawar on the west coast, where they yield an excellent building stone, and are extensively quarried (21°25'06"N; 69°49'08"E; 160 m; Khadkikar, 2004). Ancient miliolite sandstones are also quarried as a building stone near Paris.

As to geomorphological behaviour, cementation is the principal difference between most calcareous and all siliceous sand. The cement comes from the solution of $CaCO_3$ in rainwater, and its later reprecipitation. The degree of cementation depends on the proportion of calcareous sand and the ambient rainfall. Dune sands with low proportions of carbonate grains are only lightly cemented, as in the UAE and Qatar inland of the southern shores of the Persian Gulf (Williams and Walkden, 2001), and the northern Wahiba Sands in Oman, where the calcareous material was blown inland from the coast of the Arabian Sea (Warren, 1988b). But even in very dry conditions and with a moderate calcareous content, coherence develops quite rapidly, as on the Wahiba coast (Figure 3.4). In moist, warm conditions and where the sands have high proportions of carbonate, lithification begins as soon as dune movement ceases, even for quite short periods. Soils may then develop on the stabilised surface, where they serve as useful markers of the end of a sequence of deposition (Abegg et al., 2001). Because of their solubility, many calcarenites develop karst topography, including surface fissures and caves (James and Bone, 1989, RL).

Calcareous sands can be built into the full range of dune types (Louks and Ward, 2001), but because of their solubility, they seldom retain evidence of their windblown character, such as pin-stripe ripple lamination (Chapters 3 and 4; Frébourg *et al.*, 2008).

Because of their stronger cementation, calcarenites survive coastal erosion better than do siliceous coastal sands, but even so, most survivors are from the Early Pleistocene at the latest. On Lord Howe Island in the south-western Pacific Ocean, for example, an older calcarenite was dated, by a range of methods, to the early Middle Pleistocene (the penultimate major glaciation). A younger, overlying calcarenite was formed during two phases of the last major glaciation (Brooke *et al.*, 2003). Both were associated with periods with high sea levels, as are calcarenites in the Cape Province of South Africa (Butzer, 2004). A few have been reported from pre-Pleistocene strata, including those near Paris (earlier; Abegg *et al.*, 2001).

Reference

James, N.P. and Bone, Y. (1989) 'Petrogenesis of Cenozoic, temperate water calcarenites, South Australia; a model for meteoric/shallow burial diagenesis of shallow water calcite sediments', *Journal of Sedimentary Research* 59 (2): 191–203.

Chapter Twelve
Mars, Venus, Titan

Dunes were seen on Mars and Venus, even on early satellite images. As the resolution of the imagery improved, Mars was seen also to have ripples (not surprising in view of the presence of dunes), but then, at even better resolution and quite unanticipated, transverse aeolian ridges were discovered, which are unique to Mars. Even more surprising was the still later discovery of linear dunes on Titan, the biggest of Saturn's moons, in 2006.

Similarities

As regards dunes, winds are the essential similarity between Earth, Mars, Venus and Titan. The atmospheres of all four have similar atmospheric circulations: Hadley cells in the mid-latitudes and cyclonic systems nearer the poles (Chapter 4); although, unlike Earth and Mars, Venus' and Titan's Hadley cells may extend at times to their poles. Katabatic winds are stronger on Mars than on Earth, and sea-breeze-like winds sometimes occur at contrasts in thermal inertia, as on coasts on Earth (Chapter 4). All four atmospheres are also subject to 'orbital forcing', which drives long-term rhythms of climatic change (explained briefly for Earth in Chapter 10). On Mars, where the changes are more marked than on Earth, the outcome is long-term alternation of wet and dry climates (Golombeck and Bridges, 2000). As well as wetter periods, there have almost certainly been periods on Mars in which winds were more powerful (Arvidson *et al.*, 2011). Climate change on Venus is likely to have been driven more by volcanism than by orbital fluctuations, which are less variable on Venus than on Earth or Mars

Dunes: Dynamics, Morphology, History, First Edition. Andrew Warren.
© 2013 John Wiley & Sons, Ltd. Published 2013 by John Wiley & Sons, Ltd.

(Bullock and Grinspoon, 1999, RL). Titan's orbital rhythms are more complex than on Earth, Mars or Venus, given its proximity to Saturn and another of Saturn's moons, Hyperion. The apparent freshness of the Titan's dunes suggests that they are recent, which raises questions about the relationship of sand production to orbital cycles (Lunine and Lorenz, 2009).

Differences

The most conspicuous differences between these windy bodies are the physical controls on aeolian activity (covered for Earth in Chapter 1). The acceleration owing to gravity (g) on Venus is close to ours, Mars' is ~0.4, and Titan's is ~0.1 of ours. Atmospheric pressure on Venus is 90 times more than on Earth, on Mars it is 100 times lower, and on Titan it is 1.5 times higher. An early attempt to model the effects of some of these differences on the movement of sediment by the wind was made by Sagan and Bagnold (1975). Bagnold, who was 79 years old when that paper was published, was thus a pioneer in Martian aeolian geomorphology, as he had been for Earth 40 years before. Some of the speculation in the 1975 paper has been superseded, but the paper's essential message has survived. Figure 12.1 is an updated summary of its main conclusion.

A recent model puts the threshold of sand movement on the present Mars at $1.12\,\mathrm{m\,s^{-1}}$, compared with $0.26\,\mathrm{m\,s^{-1}}$ on Earth, for the same grain size (Almeida *et al.*, 2008a). This model introduced two elements that had not been included in earlier models: the viscosity of the atmosphere, which responds to temperature (among other things), and the movement of grains in a saltating cloud (against that of single particles in older models). In the model, the ratio of the dynamic (or impact) to the static (or fluid) threshold (Chapter 1) on Mars is 0.48, compared with 0.96 on Earth. In other words, saltation on the present Mars, when it occurs, can keep going at a velocity that is relatively much lower than that on Earth: Martian saltation begins at shear velocities (u_*) of ~$2.5\,\mathrm{m\,s^{-1}}$ but is sustained at ~$1.0\,\mathrm{m\,s^{-1}}$. Almeida's model also shows that saltating particles on Mars reach up to $5\,\mathrm{m}$ above the surface and can leap $120\,\mathrm{m}$ downwind (much higher and further than on Earth).

A comparison of Martian thresholds, however calculated, with what is known about contemporary Martian wind speeds, indicates that saltation on the present Mars is infrequent (Claudin and Andreotti, 2006). By one estimate, the dynamic threshold on level Martian ground is surpassed, on average, only once in five years (Fenton, 2006), and ground observations (from the Lander and Explorer vehicles) confirm the infrequency of sand movement (Greeley *et al.*, 2005). Winds off the North Polar ice cap, when funnelled down steep gullies (or chasmata), are especially fierce and may play a role in modelling the North Polar sand sea (Warner and Farmer, 2008; Figure 12.2 and later).

There are fewer data or models for Venus. They show that surface 'winds' (if wind is the word for the movement of such a dense atmosphere) may be capable of moving sand and silt, perhaps at rates as high as in Earth's deserts (Greeley

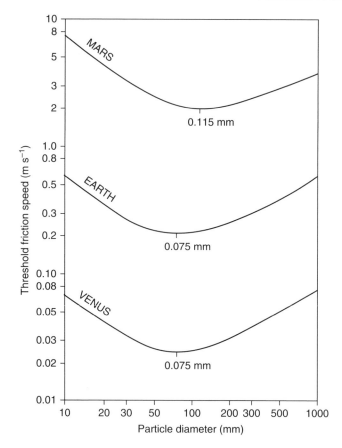

Figure 12.1   Threshold curves for particle movement on Mars, Earth and Venus (from: Greeley and Iversen, 1985). Reproduced with permission from Cambridge University Press.

and Arvidson, 1990). Reptons (Chapter 1) on Venus are up to ~13 mm in diameter, which puts them in the small gravel category (Greeley and Marshall, 1985). Saltating sand on Venus is lifted only a few centimetres above the surface (lower than on Earth), but leaps are up to a metre long. On Titan, which is less well known, the optimal wind speed for moving sand is believed to be ~0.02 m s^{-1} (at a reference height of 1 m), and the presence of dunes (shortly) hints that there are (or have recently been) winds of this speed in the equatorial regions (Tokano, 2008). Unlike Earth, Titan's equatorial winds are westerly.

Sand

Most Martian sand is dark (Figure 12.4a and 12.4d), suggesting that it is derived from iron-rich, volcanic rock, as in Iceland (Edgett and Lancaster, 1993). Its

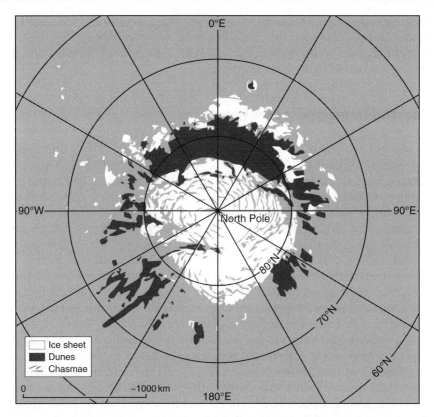

Figure 12.2   North Polar sand sea on Mars. Note the chasms in the ice cap, down which katabatic winds are strong enough to move dunes on the contemporary Mars (redrawn from: http://www.mars-dunes.org/_more/images/data/np-map.jpg).

grain size has been estimated, using images of the surface taken by the Martian Rovers, to be 0.087 ± 0.025 mm, which is coarser than most dune sands on Earth (Claudin and Andreotti, 2006).

The origin of this sand could be the more-or-less direct conversion of volcanic ejecta to dune sand (near Tharsis Mons as on Google Mars; Bridges *et al.*, 2010a), but probably a higher proportion has been produced by wind-driven abrasion: despite its thin atmosphere, the abrasive power of saltating particles on Mars is probably an 'order of magnitude greater' than on Earth, so that abrasion, over many millions of years, could have released a significant quantity of sand (Bridges *et al.*, 2005). Much more speculatively, the Martian dune sands could have been sorted by running water at some time in Mars' distant past (Weitz *et al.*, 2010). There are also some gypsum and ice dunes on Mars, as on Earth. Yet other Martian 'sands' appear to be aggregates of dust particles, held together by electrostatic forces and perhaps salts. With a lower density than other sands, the aggregate sands might be moved more easily, but it is unlikely that they would survive abrasion in saltation for long. They form 'reticulate dunes' in some craters (Chapter 4; Figure 12.4b; Bridges *et al.*, 2010a). Titan's 'sand', which is also dark, may be composed of hydrocarbon aggregates.

Ripples and Transverse Aeolian Ridges

There are ripples (Figure 12.4a; Balme *et al.*, 2008) and mega-ripples on Mars, as on Earth. Some mega-ripples were recorded (and sampled) by the Mars Opportunity Rover in 2007. Like mega-ripples on Earth, the Martian variety had coarse grains on their crests and fine material in their cores (Jerolmack *et al.*, 2006). Some of the Martian mega-ripples have wavelengths of 10–60 m and amplitudes up to 10 m (Zimbelman *et al.*, 2009), whereas the largest reported wavelength of mega-ripples on Earth is 43 m, and their amplitudes are much smaller (Chapter 2). It has been estimated that the coarse sand on Martian mega-ripples could only be moved by winds of 70 m s^{-1}. If, as also estimated, modern winds commonly reach 40 m s^{-1}, ripple activity must be very occasional indeed (Jerolmack *et al.*, 2006; although see later in relation to dunes). Ripples near Opportunity Rover are thought to have been stationary since an ancient windy epoch period (Arvidson *et al.*, 2011).

On the other hand, large ripples near Spirit Rover and the windward side of barchan dunes have been seen to move, some by 1.7 m in less than four Earth months (Sullivan *et al.*, 2008; Silvestro *et al.*, 2010b), and the tracks of Rover vehicles have been erased apparently partly by the disturbance of the surface by the wind, and partly by the deposition of dust, which was probably raised by frequent dust devils (Geissler *et al.*, 2010).

Transverse aeolian ridges (Figure 12.3) cover about 3% of the northern hemisphere (mostly south of 35°N) and 11% of the southern hemisphere (mostly north of 55°S). Three pieces of evidence suggest that they are big ripples rather than dunes: (1) the near-symmetry of their windward and leeward slopes (both at about 15°; Zimbelman, 2010); (2) their patterns of grain-size segregation, which is a very thin layer of coarse material over a core of diverse grain sizes, whose

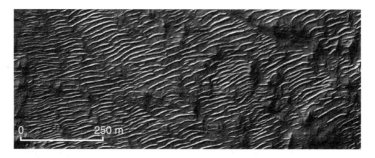

Figure 12.3   Transverse Aeolian Ridges on Mars (HiRISE ESP_024883_1525 TARs).

mean size is finer than the surface layer, as revealed by samples taken by Opportunity Rover; and (3) their frequent occurrence downwind of dune fields, which might have supplied them with the saltation-driven reptons necessary for ripple formation (as explained for earthly ripples in Chapter 2). Transverse aeolian ridges are probably inactive today, and may be relics of a former windier period (Reiss *et al.*, 2004), although their orientation is in general transverse to present winds. They may be of several ages (Berman *et al.*, 2011).

Dunes

Mars

There are dunes all over Mars (search for dunes on Google Mars; many more images are available on the web), although the proportion of the Martian surface that is covered by dunes is smaller than on Earth (Fenton and Hayward, 2010). Many of the dunes congregate on the floors of craters (Figure 12.4b) and in small numbers of other topographic 'traps' (Bourke *et al.*, 2010).

The sizes of the dunes on Mars and Venus should be calculable using L_{sat} (the saturation length), which, as explained in Chapter 1, is the main determinant of the size of Earthly dunes (Chapter 3). For both Mars and Venus, despite wide ranges in the estimation of the grain size of the dunes (a vital component in the calculation of L_{sat}), L_{drag} (the drag length that scales with L_{sat}) does appear to control the wavelength of the dunes. Martian dunes are predicted in this way to have wavelengths of $\sim 10^3$ m (as against measurements giving ~ 400–600 m) and Venusian wavelengths are predicted to be slightly greater than 10^{-1} m, clearly placing them with subaqueous ripples on Earth, a consequence of the high atmospheric density on Venus (Claudin and Andreotti, 2006).

In general, the patterns of Martian dunes, unlike their sizes, are much closer to analogues on Earth. The most immediately recognisable are barchans (Figure 12.4c), many of which have shapes in the same range as the shapes of Earth's barchans (Figure 12.4d). However, some Martian barchans look very different (Figure 12.4f), which is probably a consequence of differences in atmospheric densities, grain sizes and gravity (Parteli and Herrmann, 2007a). In many places, the horns of Martian barchans have been transformed to linear dunes (Figure 12.4d), while others, also as on Earth, join up as transverse dunes (Figure 12.4e).

Martian linear dunes are less common than on Earth (where they cover a far greater area than dunes of other kinds (Chapter 4). Some seem to have been formed by directionally bimodal wind regimes, but it is hard to fit the 'bimodal model' (Chapter 4) to all the Martian linear dunes (Figure 12.4g). Trains of barchans (Figure 12.4h) are much more common on Mars than on Earth, and these too have been modelled by Parteli and Herrmann (2007a). Star dunes and reversing dunes are rarer, perhaps because there are few areas on Mars with bimodal or multimodal annual wind directional regimes (Edgett and Blumberg, 1994;

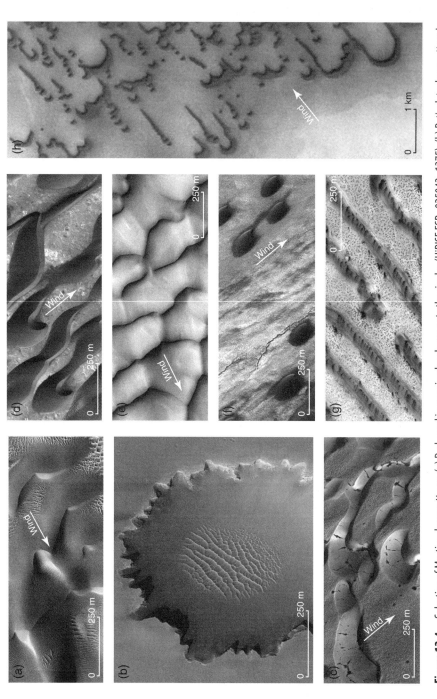

Figure 12.4 Selection of Martian dune patterns. (a) Dark sand in merging barchans; note the ripples (HiRISE ESP_025042_1375). (b) Reticulate dune pattern in Victoria Crater at Meridiani Planum (HiRISE TRA_000873_1780). (c) North polar barchans defrosting in Spring (HiRISE ESP_025042_2650). (d) Barchans and linear dunes (HiRISE ESP_023211_1940). (e) Transverse dunes (HiRISE ESP_023908_2305). (f) 'Dumpy' barchans (HiRISE ESP_018963_2650). (g) North polar linear dunes defrosting in Spring (HiRISE ESP_025044_2540). (h) Trains of barchans (much more common on Mars than Earth) (from Mars Orbiter Camera).

Bourke *et al.*, 2010; Chapter 4), but the few there are seem to be associated, as in some places on Earth, with topographic constrictions (Fenton *et al.*, 2003). Also, as on Earth, there are Martian 'forced' dunes (Chapter 5), built around hills: sand ramps, climbing dunes, falling dunes, flanking dunes and lee dunes (Chojnacki *et al.*, 2010).

The biggest question about Martian dunes is, do they still move? There is plenty of evidence of activity on their surfaces: (1) defrosting of dunes near the poles (Figure 12.4c) and possible sublimation of carbon dioxide ice nearer the equator, both of which may liberate sand to the wind (Fenton and Hayward, 2010; Hansen *et al.*, 2011); (2) small rills, perhaps fed by defrosting; (3) the shrinkage or even disappearance of some small dunes as seen on repeat imaging (Bourke *et al.*, 2008); (4) occasional signs of activity on slip faces and of the bodily movement of some dunes (as in the Meridiani Planum, at 0.4–1 m per Martian year, and perhaps elsewhere in the Martian tropics; Silvestro *et al.*, 2011); (5) modelling of flow over abrupt topography, such as the rims of craters, shows that winds are accelerated, possibly to a point where they could move sand (Bourke *et al.*, 2004c). Repeat imaging has found eight bedforms in the Endeavour Crater that have moved, although all of them also lost considerable volume (Chojnacki *et al.*, 2011).

The evidence for inactivity includes: (1) places where impact craters, landslides or yardangs have cut into dunes (Edgett and Malin, 2000); (2) signs of indurated consolidation/cementation), as simulated on Earth by a dune that had been accidentally saturated with oil; (3) the development of small 'tails' downwind of dunes, as behind nebkhas on Earth (Chapter 6) (Edgett and Blumberg, 1994; Schatz *et al.*, 2006); and (4) calculations from what is known about winds on Mars that most dunes could have moved only 1 m in *c.*7000 years (Almeida *et al.*, 2008a). It is probable that most dunes are now inactive and, like the transverse aeolian ridges (earlier), are relics of a windier past. There is some evidence of at least two generations of dune (Fenton and Hayward, 2010). However, the debate about dune movement is very active, and new evidence, both for and against dune activity, is constantly being found.

Sand Seas on Mars

A large sand sea surrounds the North Polar ice cap (Figure 12.2), to where dune formation may be driven by wind descending from the ice, as in Antarctica (Chapter 4). It is believed that, at present wind speeds, the North Polar dune fields could have formed in 50,000 years (Anderson *et al.*, 1999). There are many smaller sand seas in both hemispheres between 30 and 65°S, but the sand seas in southern high latitudes are generally much smaller than those in the northern. Most of the dune fields in the south are held in craters rather like the sand seas in tectonic basins on Earth (Chapter 8).

As on Earth (Chapter 8), but at a more global scale, sand is transferred between sand seas and dune fields on Mars. In the high northern latitudes, winds take sand towards the Polar dune fields (where it meets the katabatic winds of the ice cap). In the lower northern latitudes, sand is blown to the southern hemisphere.

Venus

On Venus, most 'dunes' are at low elevation, where atmospheric pressure is greatest (Marshall and Greeley, 1992). Only two dune fields have been identified, in which the dunes appear to be composed of volcanic sand. The dunes are spaced at ~0.5 km but have low amplitudes by comparison with dunes on Earth or Mars (Bourke *et al.*, 2010). The dunes appear to degrade over lengthy periods of time, as sand is blown away (Bondarenko and Head, 2009). There are also fields of 'micro-dunes' (wavelength ~0.5 m), and these have been reproduced in a Venusian wind tunnel (Greeley *et al.*, 1984c). No detectable movement occurred in any of these dunes over an eight-month period (Greeley and Marshall, 1985). A possible aeolian oddity on Venus may be 'aeolian' flat beds, which were developed in the Venus wind tunnel at Ames, California, at flows over $1.5\,m\,s^{-1}$ (Greeley *et al.*, 1984c). Flat beds occur under fast-flowing water on Earth, but subaerial flat beds are unknown, although they are a theoretical possibility in high winds (Momiji and Warren, 2000).

Titan

Linear dunes cover 12.5% of Titan's surface, amounting to a considerably higher proportion of its area than of Earth or Mars, and a total area greater than that of the United States (Le Gall *et al.*, 2011). Their spacing is of the order of 4 km. As far as is known, most dunes occur near the equator (Figure 12.5), probably because the lower latitudes are drier than poleward areas ('drier', that is, in terms of methane rather than of water on Earth; Mitchell, 2008). The configuration of the dunes suggests a generally westerly wind regime, and wind speeds sufficient to move sand in present conditions. The dunes are all linear and are probably

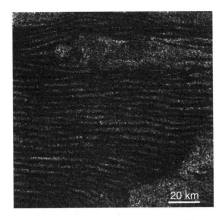

Figure 12.5   Linear dunes on Titan (Bourke *et al.*, 2010; public domain/Elsevier).

aligned to the resultant in a seasonally variable wind regime, created by fluctuation of the Hadley circulation (despite some doubts about wind directions; Tokano, 2010). Alternatively, cohesion of the hydrocarbon sands might inhibit mobility, allowing the linear dunes to develop downwind in an almost unimodal wind regime, as has been proposed for some linear dunes on Earth (Rubin and Hesp, 2009; Chapter 4). The dunes bifurcate round highlands, as on Earth (Radebaugh *et al.*, 2010). In one area of Titan, the mean spacing of the linear dunes is 2.3–3.3 km, and their heights are 45–80 m, rather larger than 'sand ridges' but smaller than 'large linear dunes' on Earth (Chapter 4; Neish *et al.*, 2010).

Reference

Bullock, M.A. and Grinspoon, D.H. (1999) 'Global climate change on Venus', *Scientific American* March 280 (3): 50–57.

Part Four
Care

Chapter Thirteen
Local, Short-Term Care ($<1000\,m^2$; <10 years)

Dunes in Deserts

Moving sand and dunes have threatened desert oases for thousands of years. In El Hasa oasis in eastern Saudi Arabia, it has been estimated that $3.5\,km^2$ of palm orchards were lost to encroaching dunes over 50 years (25°25′59″N; 49°42′56″E; 4 km; Hidore and Albokhair, 1982).

Folk science

In the tropical and subtropical deserts of the Old World, the time-tested technology for protecting gardens and homes against the encroachment of sand is a fence of date-palm fronds. The location and alignment of the fences are informed by an intimate knowledge of where the sand comes from, what it might threaten and to where it might be diverted (as in the system of angled fences around a palm orchard near Kharga in Egypt: 27°27′21″N; 30°39′90″N; 180 m). Although evolved independently, this technology shares many features with the scientifically tested techniques that were later deployed. For example, a controlled experiment has shown that double fences, like some used in folk systems, are more cost-effective than dispersed ones, because they need less maintenance (Miller *et al.*, 2001; later).

Similar kinds of understanding guided two other folk defences against blowing sand and dunes. The first was the discovery of the platelayers (trackmen) on a Peruvian railway in the early 20th century. By spreading loose pebbles and grit

Dunes: Dynamics, Morphology, History, First Edition. Andrew Warren.
© 2013 John Wiley & Sons, Ltd. Published 2013 by John Wiley & Sons, Ltd.

thinly on the surface of approaching barchans, they increased near-ground turbulence, which in turn induced the dunes to shrink and disperse, leaving the sand to saltate harmlessly over the track (16°41′S; 71°51′W; 3 km; Bailey, 1917; Bagnold, 1941, p. 180). This system, too, was validated scientifically much later by Logie (1982). The second was the work of a celebrated 'sand engineer', Belkacem, in El Oued Souf in Algeria, to protect roads in the 1940s and 1950s (33°30′N; 6°47′E; 4 km; Mason, 1951). He knew just where to place protective walls of sand capped by a layer of clay. This, too, now has a more scientifically based, but arguably no more effective, equivalent (Zhang Chunlai *et al.*, 2007a).

New approaches

Folk methods are still the only option for most rural desert people, but as the commercial value of the installations at threat has increased, so have the sums for protection and, in turn, research. The reality and severity of the problems that dunes present to expensive installations are illustrated by the burial of a new highway near Kharga in Egypt (25°23′09″N; 30°27′55″E; 5 km).

Since the first oil wells were sunk in the South East Asian deserts in 1913, commercially funded research has produced some principles for sand control (summarised in Cooke *et al.*, 1982, p. 277). The first of these is avoidance, a real option in the vast open spaces of many deserts, and for structures like roads, which are not wholly constrained by fixed points like oil or water wells or settlements. The ease with which the older roads around Kharga in Egypt have avoided streams of sand and dunes is a case in point (25°25′37″N; 30°37′31″E; 300 m). Avoidance is easiest where the origin of the sand and the strength and direction of the winds that blow it are identified, as in Tabuk in north-western Saudi Arabia (Figure 13.1).

If avoidance is impossible, and a road or a railway are still in the path of moving sand or dunes, the route should cross a gentle slope rising in the downwind direction, where sand movement is accelerated. Twists or pockets, as in standard designs for cuttings or bridges, should be redesigned to encourage the wind to empty them of sand. The strategy can be adapted to areas with little slope by building one, in other words running the route atop an embankment with smooth side slopes of between 1:5 and 1:6. But, embankments with these specifications hold back only 1–2 m high dunes, no bigger (Watson, 1990).

In some countries, like Algeria, anti-terrorist regulations require pipelines to be buried up to 2 m down. Trenches in dunes are the cheapest to excavate, but is the regulation depth enough to avoid exposure by erosion, which might leave the pipes suspended between dunes? The model of Momiji and Warren (2000), introduced in Chapter 3, was built to test that likelihood and showed that excavation by the wind was only likely in high and persistent winds, which are very rare (Figure 3.3).

The next resort, if the other two are not possible, is defence. The checkerboard patterns of low fences within which bushes are planted, which work well in

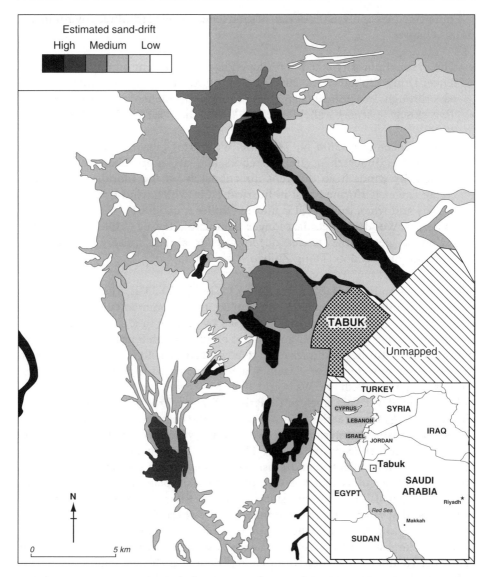

Figure 13.1   Potential origins of blowing sand in the vicinity of Tabuk in north-western Saudi Arabia (Jones *et al.*, 1986).

semi-arid parts of China (later), can be overwhelmed by sand in very arid areas, as on the trans-Taklamakan highway (Lei JiaQiang *et al.*, 2002) and Mauritanian Route 3 (built as a famine-relief road) in the south-east of the country (16°15′23″N; 8°08′01″E; 500 m; located on Figure 10.11). Inert fences are then the only viable option for protection from sand encroachment (Jensen and Hajej, 2001).

Four principles apply in the design of sand fences in all of the three environments of this chapter:

- The optimal porosity is 0.3–0.6 (depending to an extent on the height of the fence; Dong Zhibao *et al.*, 2006b). More porosity allows too much sand to pass through; less porosity can create enough turbulence to destroy the fence.
- Two closely spaced parallel fences are better than one (earlier) (Miller *et al.*, 2001).
- Zigzag fences (39°52'06"N; 74°05'00"W; 250 m) create wider, lower, more manageable dunes than straight fences and are more likely to trap sand brought by winds from different directions, as on the Florida coast, where zigzag fences performed best in hurricanes, in which wind directions change quickly and when most sand is moved (Miller *et al.*, 2001).
- Materials. Wooden fences last longer than textile fences (Miller *et al.*, 2001). Nylon or 'geo-textile' fences decompose or abrade quickly but can give temporary protection (Dong Zhibao *et al.*, 2004a).

Where irrigation is affordable (as at 24°13'11"N; 54°50'25"E; 5 km), the planting of grasses is not recommended (quite unlike the recommendations for coastal dunes, later), because they are quickly buried by moving sand, have roots that are too shallow to tap groundwater and do not retain enough foliage to act as perennial sand traps. Bushes and trees need to be spaced so as to reduce rather than to increase sand movement (at certain spacings, the wind is accelerated between clumps; Chapter 6; Ash and Wasson, 1983). Many tree and bush species tolerate only occasional watering, such as Tamarisks, Acacias, Prosopises, Casuarinas and Eucalypts. At El Hasa (earlier), Tamarisks were the best choice, because they root deeply, do not need watering after two to three years and are tolerant of quite saline groundwater (Stevens, 1974). There is a large literature to help in the choice of the most effective and most ecologically sound species (Wilkie, 2002, RL). The consensus is that the best place to start planting (or spraying with chemicals, later) is the windward slope of a dune, where the plants will trap some of the sand that might have reached the slip face; where they will not be buried by an advancing slip face; and where their roots are closer to moisture deep in the dune. The largest literature and most extensive experience in protection against sand encroachment now come from China (Zhao Wenzhi *et al.*, 2008).

Some of the plantations designed to protect against blowing sand and dunes are vast. At El Hasa, planting began in the early 1960s on a sandy area about 5 km wide (in the direction of the wind). In the first phase alone, five million trees were planted in 500 ha of moving sand (coordinates given earlier; Achtnich and Homeyer, 1980). The trees were irrigated for the first five years to allow roots to reach the water table. In the Chinese deserts, there is a major programme of similar kinds of irrigated planting to protect railways (whose design is discussed later). Irrigated plantations are much cheaper close to canals, as along the lengthy Indira Gandhi Canal in Rajasthan in India (27°59'27"N; 72°30'15"E; 500 m).

Another approach is to reduce the rate of the entrainment of sand. Spraying with water lays sand, and is affordable in towns and on dumps of fine material near mines and power stations but is too expensive for bigger or remoter areas, and its effect is very short-lived. A somewhat cheaper and longer-lasting technique is to spread coherent soil or gravel on the surface, at a density that must be related to the purpose in mind. If the purpose is simply to disperse a dune that might threaten to engulf some worthy object, without concern for the downwind change in the discharge of sand (as on the early Peruvian railway, earlier), then the concentration will depend on local wind conditions and would have to be arrived at by trial and error, or better by some dedicated wind-tunnel or field research, in the manner of Logie (1982). If the purpose is to prevent the surface from yielding sand, a cover of ~30%, is recommended, as for the protection of the ancient Buddhist art in the Mogao Grottoes in very dry north-western China, the earliest of which dates from 366 CE (40°02′27″N; 94°48′28″E; 4 km; Liu Benli et al., 2011). The threat at the Grottoes is blowing sand rather than moving dunes, but if dunes threaten to cover a protected area, as at El Hasa, gravel-spreading would give only temporary protection, unless it were to cover a very large area, and be very costly (coordinates given earlier; Achtnich and Homeyer, 1980).

Some valuable installations, such as roads, can be protected by applying stabilising substances on any surrounding bare sand. Among the pioneers in this approach were Soviet chemists (Akhmedov et al., 1969). There are now many substances on the market for this purpose. Asphalt and other oil-derived stabilisers are the most widely used (Dong ZhiBao et al., 2004a); polyvinyl acetate and related polymers have been proven to be very effective (Homauoni and Yasrobi, 2011). Some stabilisers, though effective, are unsightly or toxic. Specially designed chemicals (including ones based on latex, sodium silicate), have varying brittleness, hardness, viscosity (and hence penetrability), permeability and resistance to UV light or oxidation (Han Zhiwen et al., 2007). Spraying in strips permits more area to be covered with the same volume of stabiliser and, carefully done, can be effective. Sprays may be cheaper than they were but are still only cost-effective for the protection of important facilities or routes. This has not prevented well over 50 km² being treated in the former USSR. In Saudi Arabia, one company alone sprayed 35 km² between 1978 and 1983 (Watson, 1990). Sprays can also be used to provide protection from moving dunes. As with planting (earlier) the best place to spray a dune is its upwind slope.

Many other tactics and devices have been used. One is to dig pits to intercept saltating sand, but they need to be constantly emptied. Another is to accelerate flow by the skilled arrangement of impermeable fences (earlier, as a folk method). Sand-trapping fences can be used in these environments, as they are on beaches and semi-arid areas (later). Yet none of these methods is totally effective; they seldom do more than alleviate or postpone the problem (Stipho, 1992).

There are two last resorts. The penultimate option is to remove dunes. A principle from Chapter 3 is worth recalling: small dunes travel more quickly than large ones and thus should be the first targets. Removal can be achieved with the aid of the wind itself, as when dunes are reshaped to encourage dispersal or sand trapping, but the usual approach is more brutal: bulk removal, which at its most expensive involves bull-dozing the dune, a practice that is very obvious in Sinai at 30°53′53″N; 34°04′00″E; 3 km. But taking out a dune is seldom a one-off operation: more dunes usually follow. The same can be said of the last resort, sweeping sand, as is done on many of the highways in Arabia, but there will always be more to sweep.

Stabilised Dunes in Semi-Arid Areas

On the desert margins, the principles of dune fixation are much the same as those in temperate coastal dunes (shortly), albeit with other sand-fixing species. The technical solutions, however, may differ.

As on coastal dunes, sand fences are a widely used technique but not, as in very arid areas, simply to keep sand from drifting, but also to 'nurse' plantations before they are themselves able to act as a protection. In these cases, it has been found that a rectangular arrangement (a 'checkerboard') is best, as seen in the protected zone round a road and railway in central China (37°27′N; 104°58′E; 1.6 km; Zhang Kebin and Zhao Kaigo, 1989). Checkerboards are also used in the drier parts of China, but they must be irrigated. The optimal size of the squares is 2 × 2 m. Bigger ones provide little protection; smaller ones give no more protection and are more expensive (Qu Jianjun et al., 2007e). Even fences only 10–20 cm high can be effective. The fences also encourage dust to settle, to the further benefit of soils and plants (Li Yulin et al., 2009). Where labour is plentiful and brushwood in short supply, the fences may be made of clay (as in El Oued Souf, earlier). There are now many techniques for encouraging plants to establish, such as planting them in earthenware pots, which help to retain moisture and nutrients. New plantings invariably need to be watered for a few years before they establish, and may need to be fenced against domestic stock and the collectors of firewood, browse or hay. If the site to be protected in an area with moderate rainfall, and if the surface is stabilised, as with a mulch, planting may be unnecessary, because there may already be enough seeds in the soil, or in the wind, to ensure that plants will be re-established, as in parts of southern Texas (Fulbright et al., 2006).

One method of sand control is the inoculation of the soil with crust-forming cyanobacteria. An early experiment showed a marked increase in the thickness and biological diversity of the treated crust, and colonisation by higher plants three years after inoculation (McKenna-Neuman et al., 1996). Aerial seeding can be more extensive, and is sometimes successful, but drifting sand may bury emerging seedlings (Zhang Chengyi et al., 2002).

Coastal Dunes

It has been alleged that in the Culbin Sands in Morayshire, Forvie in Aberdeenshire, the Gower in South Wales, Legé in the Landes of south-western France, Tved on the south-western coast of Denmark and Berrien County in south-western Michigan, whole villages were buried by reactivated coastal dunes, sometimes initiated by a storm (Blanchard, 1926; Reber, 1928; Edlin, 1976; Lees, 1982b; Skarregaard, 1989; Robertson-Rintoul and Ritchie, 1990). The stories may have been overegged, but they hold some truths. At Tved, the church alone survived. At Mimizan in the Landes, sand buried 16 m of a church tower in the 17th century. The depth is still marked on the tower (Figure 7.1; Fenley, 1948; Clarke *et al.*, 2002), and there is potential for bigger disasters. If the dunes on the Dutch coast (52°15′N; 04°26′E; 4 km) were to be breached, which is a real possibility, an area with well over a million people might be flooded. Coastal dunes provide dubious protection for many other highly productive areas, including much of the southern Atlantic coast of the USA.

Coastal dunes cover huge areas and have many values. In Britain, which is not especially well endowed, 4% of the land area is covered by coastal dunes. A quarter of the southern Australian coast (Short, 1988a), virtually the whole of the Belgian coast, 80% of the Dutch coast and most of the German and Polish coasts, and vast stretches of coast in south-western and southern France, the Ukraine, Romania, Israel, northern Egypt, eastern India and the eastern and western coasts of the United States are backed by dunes. In many parts of the world, dunes hold massive reservoirs that serve domestic water supplies. In the 1980s, the Dutch dunes supplied four million people with $40 \times 10^6 \, \text{m}^3 \text{yr}^{-1}$ of water. The aquifers were recharged artificially; the sand acted as a filter bed (van der Meulen and Jungerius, 1989a). Commercial forestry covers vast areas of coastal dunes (44°19′N; 03°03′W; 900 m). In northern New South Wales and southern Queensland, beach sands, rich in zircon and rutile, have been taken from coastal dunes since the 1940s; perhaps 5% of a 1000 km stretch of coast will eventually be mined, with the potential for massive devastation (Morley, 1981). Coastal dunes are quarried for building sand in New Zealand and in many other places (Hilton, 1994). Coastal dunes are the most intensely used of recreational resources, and hence the most familiar of landforms. Golf was invented there, bunkers (blowouts) and all (natural and contrived blowouts can be compared at 56°21′12″N; 2°48′55″W, 550 m, on the Old Course at St Andrew's). The ecosystems of coastal dunes are described in many textbooks (Maun, 2009, RL), and they attract thousands of students on field courses. They are the most accessible of the few remaining 'natural' habitats available for ecological research; they are a valuable wildlife habitat, with distinct and rare species; and they render many other ecological services (shortly).

At the small scale, care is a matter of applying aeolian geomorphology and plant ecology, with few complications. At this scale, the prime distinction is between the active front-line dunes (embryo dunes and fore-dunes) and the less active dunes inland. Two characteristics of front-line dunes, as described in Chapter 7, are important to their care. First, the dunes and the beach are a closely coupled system. Second, storm waves do the most damage. Care must therefore focus on trapping sand on the dunes as a defence against erosion in storms. At its crudest, this may mean bulldozers (Mitasova *et al.*, 2005), but, emergencies apart, there are many subtler and cheaper methods.

Sand fences are the most common technique for fixing coastal dunes. Fences defend vast stretches of coast, for example, 82% of the New Jersey shore (Grafals-Soto and Nordstrom, 2009). Apart from their role in trapping sand (as in the other environments of this chapter), they prevent access to people and livestock. This allows plants to colonise, which adds further protection, provides a more pleasing backdrop to the beach and protects biodiversity. The species that are planted in the areas protected by fences need to be able to survive salt spray and burial by sand. The grass *Ammophila arenaria* (marram grass) is the staple (Chapter 6). A sterile hybrid, *Calammophila baltica* may be better but has yet to be extensively adopted (Arens *et al.*, 2001b). Marram has been pivotal to the history of dune fixation. Special regulations covered its planting in fourteenth-century Holland (van der Maarel, 1979), and it was folk knowledge of marram or oyat that Brémontier built upon in his groundbreaking systems for fixing the fore-dunes in the Landes of south-western France in the eighteenth century. After nearly two centuries of systematic dune fixation, there is now a weighty body of knowledge on how, where and when to plant (Houston *et al.*, 2001).

If there are valuable installations to protect, there may be intense pressure to control wave attack on the dunes. Pressure can be countered with the valid claim that attacks by waves usually take sand back to the beach, where it may reduce the severity of further attack (Chapter 7). An obvious strategy for harnessing this process is to restrict the building of valuable installations close to the shore. If this fails, as it often has, attack can be countered by the mechanical modification of the outer dune slope and nearby beach or, at worst, 'hard engineering', as with concrete barriers. Engineering models are available that can predict the best configuration for loose sandy and concrete slopes (summarised in Sherman and Bauer, 1993b).

Beach 'nourishment' (taking sand, from older dune sand inland or from offshore, and adding it to the beach) is now widely deployed in attempts to maintain coastal systems (beaches and dunes). The longer-term implications of beach nourishment are discussed in the next chapter, but even in the short term, which is the usual target of nourishment, the consequences are not all benign. Problems arise if the 'nourishing' sediment differs in grain size from the existing beach or dune. On one part of the Portuguese coast, nourishment with sediment that was much finer than that on the beach or the dunes resulted in its rapid loss by erosion, both windblown and by runoff. The sediments had been dredged close to the shore, which itself seemed to induce greater coastal erosion down-drift (Matias *et al.*, 2005). In Spain, a beach-nourishment scheme also used fine sand, and this too blew onto the dunes, where it buried vegetation and some beach houses (Marqués *et al.*, 2001). These, of course, are not insuperable problems (although better-judged designs may be more costly).

The longer-term consequences of beach nourishment, and the other techniques discussed in this chapter, are issues of 'sustainability' (Chapter 14).

References

Maun, M.M. (2009) *The Biology of Coastal Sand Dunes*. Oxford: Oxford University Press.
Wilkie, M.L. (2002) 'From dune to forest: biological diversity in plantations established to control drifting sand', *Unasylva (FAO)* 53 (209): 64–68.

Chapter Fourteen
Sustainability ($>100,000 \text{ m}^2$; >10 years)

The sustainable management of dunes (or any other natural geomorphological system) is much more complicated, uncertain and controversial than management at the small scale. At the small scale, say of a few hundred metres of beach and dunes, or with a few hectares of sandy land, the natural system can be moderately well modelled; learning can be based either on trial and error or on tinkering with simple models; a failure in management is quickly recognised and as quickly rectified; few people are involved and have a limited range of value systems, so that disagreements can usually be resolved. The management of desert dunes can operate within such a limited temporal and spatial frame, but sustainability is a real concern in the management of most of the dunes described in this book.

Constraints

Complexity

Following Schumm and Lichty (1965, RL; Introduction), increasing scale increases complexity. There are: more wind regimes; sets of dry and wet years; other physical processes like water erosion, which interfere with the workings of the wind; different plant communities, which offer different types of protection; and sedimentary pathways that are longer and more varied. Above all, there are more feedbacks and lags. In 'The Techno-Human Condition', Allenby and Sarewitz (2011; p. 28, RL) argue that decision-making becomes extremely

Dunes: Dynamics, Morphology, History, First Edition. Andrew Warren.
© 2013 John Wiley & Sons, Ltd. Published 2013 by John Wiley & Sons, Ltd.

complex at these higher scales, where systems are 'integrated in ways that can never be fully understood'. Hence, flexibility is a necessity.

In coastal dunes, complexity begins with the supply of sand, which, as scale increases, is delivered over greater and greater distances and periods of time (Chapter 11). Remote sources, events and policies can affect the pathways followed by sand to the dunes in question. When sediment input into the Nile is reduced, say by the building of a dam, sand supply to Israeli coastal dunes is eventually reduced, and the coastline is in greater danger of receding (Zviely et al., 2007).

Even at an intermediate scale, the delivery of sand to coastal dunes is complex enough to cause problems. Coastal defence can restrict the flow of sand from eroding cliffs, and this can endanger dunes at a distant site. Fourteen per cent of the Californian coast between Marin County and the Mexican border had been armoured against erosion in the first decade of the 21st century (one form of armouring is at 33°11'15"N; 117°22'43"W; 80m). But the cliffs on an unarmoured Oceanside section of this coast (32°42'30"N; 117°15'18"W; 80m) contribute 12% of the sand in downdrift beaches and dunes. Armouring on that stretch of coast, or on others, might therefore seriously threaten beaches and dunes downdrift (Runyan and Griggs, 2003). Jetties and groynes (barriers across beaches intended to control long-shore drift) may interfere with the supply system to a coastal dune, as near Biddeford in Maine (43°27'39"N; 70°22'22"E; 2km; Kelley et al., 2005). On the southern coast of South Africa, well-meaning projects stabilised coastal dune fields that had, when active, carried sand from one beach to another across headlands (known as 'headland bypass dunes' as at Cape Agulhas 34°47'S; 19°59'E; 14km; Cock and Burkinshaw, 1996). After the stabilisation, the downdrift beaches and dunes were starved of sand, and some were eroded. On the southern Brazilian coast, the devegetation of another set of headland bypass dunes has been sanctioned in the hope of maintaining the supply of sand to other downdrift beaches and dunes (27°27'S; 48°23'S; 6km; Boeyinga et al., 2010). On the New Jersey coast, dune fixation at one point withheld sand from downwind dunes (Nordstrom and Gares, 1990). Centuries of destructive and constructive interference have also wrought huge changes in the Manawatu dunes in New Zealand (Hesp, 2001) and in the Netherlands (Nordstrom and Arens, 1998).

Another facet of complexity is the widening of concern to spheres other than the geomorphological. One of these concerns, which plays a prominent part in the care of coastal dunes, is the long-term maintenance of plant and animal communities. This is an issue in which the Dutch have taken the lead: a national tradition of interest in coastal dune ecosystems has been behind a policy of encouraging bare sand in coastal dunes (in the form of blowouts, Chapter 6), which favours species such as the sand lizard (*Lacerta agilis*). The results can be seen at 52°08'51"N; 4°20'37"E; 2km (Arens and Geelen, 2006).

The 'human' part of the Anthony–Sarewitz argument is seen in the complexity of socio-political and management systems within which the care for coastal dunes is situated. At the wider scale, a coastal dune is no longer the prerogative of engineers, who must then liaise with other groups in this expanding system, each with its own standards and objectives. The more interests there are, the more complex becomes their interaction. In many cases, the sources of sand for a particular dune are under a different political control than that of the dune itself. The whole system is further complicated as political authorities multiply. Regulations and finance, be they from State/Federal/National/European Union or from other regional groupings/United Nations agencies, can clash and may bring conflicts of loyalty, as in northern France (Meur *et al.*, 1992).

Uncertainty

Uncertainty multiplies as complexity increases. Scientific uncertainty, for a start, increases with scale, as seen in the greater uncertainties in Chapter 4 than in Chapter 3.

At large scales, it is institutions that are the managers, and institutions create their own uncertainties, as they bend first to one interest group and then another, and as their legislative foundations and funding waver, causing them to become more self-defensive, evasive, vague or contradictory about their goals and achievements. There has been a healthy debate about rates of accelerated erosion, by wind and water, in the United States. Trimble and Crosson (2000a) maintained that the United States Department of Agriculture had been basing its strategies on data from small plots but that rates for bigger areas, which were not accurately known, were likely to be much less. Others disagreed (Trimble and Crosson, 2000b).

Environmental change

Environmental change is another huge uncertainty, but also introduces new issues. All of the processes discussed in this chapter would be affected if global temperatures were to rise, especially if wind speeds, rainfall or evaporation were to follow suit.

Of windblown landforms, coastal dunes are unique in their vulnerability to the rise of sea level which has already risen and will fairly certainly accelerate (Figure 14.1). As explained in Chapter 7, most coastal dunes are already relics of changes in sea level which have been, in general, rising in the last 10,000 years. The release of sand and its movement to coastal dunes that this triggered decelerated as sea levels stabilised. Many coastal dunes now lose more sand than they gain. A worldwide survey showed sandy coasts to be already in retreat almost everywhere (Pernetta, 1994, RL). In the Netherlands in the 1990s, 38% of the coast was regressing, 39% was stable, and only 23% was prograding (Arens and Wiersma, 1994a).

The consensus is that sea level will rise at a rate of between 3 and 6 mm yr^{-1} in the next few decades. All coastal dunes would be affected if this were to happen but in different ways. Dunes that are isolated from sand supplies are the most vulnerable. Storm waves would erode them more often, and their recovery would

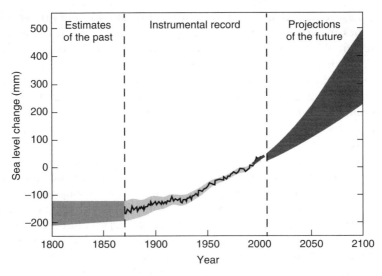

Figure 14.1   Sea level, then, now and soon (Bindoff *et al.*, 2007).

be slower. On the other hand, many downdrift beaches and dunes, particularly those at the ends of sediment transport systems, may receive more sand (Psuty and Silveira, 2010). If the new sand did not bury marram (Chapter 7) (whose growth might at the same time have been stimulated by increased CO_2, temperature and rainfall), more extensive vegetated dune systems would be the result. But if the new supply of sand reached some threshold, it might bury marram and set dunes free (Carter, 1991). This is what seems to have happened on many coasts at the end of the Pleistocene, perhaps after an abrupt climate change, and is still happening on some coasts.

Coastal dunes are likely to suffer other effects of climate change. There may be more salt spray in stormier conditions, killing plants. Erosion may lower water tables, creating drier conditions and also killing plants (Pernetta, 1994, RL). Elsewhere, water tables might rise along with the rise in sea level. New climates might desiccate or drown existing plant species (by analogy with the Holocene history of coastal dune fields – Chapter 11).

Sustainability

Coastal dunes

Sustainability is best introduced by looking at coastal dunes, where it can be expressed as an equation:

$$\textit{sand in} \text{ should equal } \textit{sand out}.$$

The formula could, perhaps should, be applied by law to sand quarrying in the coastal dunes in New Zealand. On dunes with an active input of sand, quarry companies should remove no more than is being added to the dunes from the beach. On ancient stabilised dunes, the amount taken should be insignificant compared with the total reserve of sand. But in New Zealand, coastal dunes are not being adequately surveyed for the data required by New Zealand's Assessment for Environmental Effects in its aims properly to control quarrying (Hilton, 1994).

Beach nourishment (Chapter 13) balances the sustainability equation by manipulating the input side of the coastal sedimentary budget. Without nourishment, some coastal dune systems and habitats would survive, and some would increase in area, but the landform type, as a whole, would certainly diminish. But as good as well-designed beach nourishment may be for coastal dune habitats in the short term, it is rarely sustainable. It is a commitment to endless investment, the rate of which would increase if the property owners whom it defended were made to feel secure enough to themselves invest more, and so demand yet more protection. Nourishment also means a continuing or more probably increasing input of energy (because of the need for accelerated nourishment as sea levels rise and as supplies of suitable sand diminish), and energy is becoming dearer.

Beach nourishment is not the only intervention that might affect the sustainability of coastal dunes. Another is the redistribution of sand following storm damage, for example of overwash deposits (which are taken through breaches in the dune by seas raised by storms). Where it has closed roads or beach houses, this sand is usually removed and taken to build dunes as defences against coastal erosion. Mathematical modelling of the long-term effects of one of this type of intervention has shown that they produce narrower, steeper fore-dunes, which may actually exacerbate overtopping by the waves of very intense storms and, as a whole, create a much more erratic behaviour of the beach–dune system in the long term (Magliocca et al., 2011).

One strategy for the sustainable care of coastal dunes, in complete contrast to the rebuilding of fore-dunes, is 'managed retreat', which is now seen, even by many engineers, as not only a better way to conserve habitats, but is also a better way of defending the coast, and at a lower long-term cost than hard engineering or beach nourishment. One form of managed retreat is the experimental breach in an outer dune barrier in the Netherlands (the ultimate defence against the North Sea), to create a 'slufter' or wet dune slack that is invaded periodically by the tide, a habitat that has been endangered by hard engineering (a slufter is at 52°41′03″N; 04°38′13″E 800 m; Arens et al., 2001b). Brave as these initiatives are, they do little to protect the polders and cities behind the dunes, unless part of a much more elaborate and extensive plan.

Stabilised inland dunes

The expansion of the area of active windblown sand in now stabilised dune systems would accelerate in the warmer world that is forecast by most climate models. Many semi-arid parts of the world would become windier and some drier. There have been several studies of 'sandification' or 'sandy desertification' in the semi-arid parts of China, based on sequential remote sensing. They show periods both of expansion and of the contraction of bare sand in the recent past, in which land use seems to have been a stronger factor than the climate (Guo JiAn et al., 2010). However, if droughts became more severe or more frequent, as some expect, the task of keeping sands stabilised might become more difficult.

On the Great Plains of North America, forecasts of Global Climatic Models have been combined with a climatic index of dune activity (Mangan et al., 2004). The outcome suggests that if droughts were to be as severe as some in the last few thousand years (Chapter 10), as they may well be, patchy dune reactivation would certainly occur, even without any change in the intensity of land use. A similar approach has predicted an alarming expansion of dune activity in the Kalahari in southern Africa (Thomas et al., 2005a), although a newer study, using different models both of climate change and of the relationship between plants and moving sand, found it unlikely that more sand would be mobilised in either the Kalahari or in Australia in the medium-term future (Ashkenazy et al., 2012).

This chapter has shown that there are ways to lessen the problems that climate change may bring, but at what cost? Another route, which now seems less and less likely to be taken in time, is to mitigate climate change by controlling gaseous emissions.

References

Allenby, B.R. and Sarewitz, D. (2011) 'We've made a world we cannot control', *New Scientist* 14 May (2812): 29.

Bindoff, N.L. et al. (2007) 'Observations: oceanic climate change and sea level', in S. Solomon et al. (eds), *Climate Change: The Physical Science Basis*. Cambridge: Cambridge University Press, pp. 385–428.

Pernetta, J. (ed.) (1994) *Impacts of Climate Change on Ecosystems and Species: Implications for Protected Areas*. Gland: IUCN, 108pp.

Schumm, S.A. and Lichty, R.W. (1965) 'Time space and causality in geomorphology', *American Journal of Science* 263 (2): 110–119.

Zvively, D., Kit, E. and Klein, M. (2007) 'Longshore sand transport estimates along the Mediterranean coast of Israel in the Holocene', *Marine Geology* 238 (1–4): 61–73.

Index

Page numbers for main entries are in bold; page numbers referring to figures are in italics; features with Google Earth coordinates are indicated with an asterisk.

Dunes: Dynamics, Morphology, History, First Edition. Andrew Warren.
© 2013 John Wiley & Sons, Ltd. Published 2013 by John Wiley & Sons, Ltd.